U.S. Marines in Afghanistan, 2010–2014
Anthology and Annotated Bibliography

U.S. Marines in the Global War on Terrorism

Compiled by
Paul Westermeyer
with Christopher N. Blaker

History Division
United States Marine Corps
Quantico, Virginia
2017

Other Publications in the Series
U.S. Marines in the Global War on Terrorism

U.S. Marines in Iraq, 2003: Anthology and Annotated Bibliography

U.S. Marines in Iraq, 2003: Basrah, Baghdad and Beyond

Det One: U.S. Marine Corps U.S. Special Operations Command Detachment, 2003–2006

U.S. Marines in Iraq, 2004–2008: Anthology and Annotated Bibliography

U.S. Marines in Iraq, 2004–2005: Into the Fray

U.S. Marines in Iraq, 2003: Combat Service Support During Operation Iraqi Freedom

U.S. Marines in Afghanistan, 2001–2002: From the Sea

U.S. Marines in Afghanistan, 2001–2009: Anthology and Annotated Bibliography

Library of Congress Cataloging-in-Publication Data

Names: Westermeyer, Paul W., compiler. | Blaker, Christopher N., compiler. | United States. Marine Corps. History and Museums Division. Historical Branch, sponsoring body.
Title: U.S. Marines in Afghanistan, 2010/2014 : anthology and annotated bibliography / compiled by Paul Westermeyer with Christopher N. Blaker.
Description: Quantico, Virginia : History Division, United States Marine Corps, [2017] | Series: U.S. Marines in the global war on terrorism | Includes bibliographical references and index.
Identifiers: LCCN 2017001961
Subjects: LCSH: Afghan War, 2001- | United States. Marine Corps--Afghanistan. | United States. Marine Corps--History--20th century. | United States. Marine Corps--Afghanistan--Bibliography. | United States. Marine Corps--History--20th century--Bibliography.
Classification: LCC DS371.412 .W475 2017 Z1361.D4 | DDC 016.958104745--dc23

Table of Contents

Part III: Every Marine a Rifleman, 2012

Part IV: Advising and Nation Building, 2013

Part V: Exiting Afghanistan, 2014

Part VI: Looking Back

Appendices

Foreword

This volume presents a collection of 21 articles, interviews, and speeches describing many aspects of the U.S. Marine Corps' participation in Operation Enduring Freedom from 2010 to 2014. This work is intended to serve as a general overview and provisional reference to inform both Marines and the general public until the U.S. Marine Corps History Division completes monographs covering major Marine Corps operations during the campaign. The accompanying annotated bibliography provides a detailed look at selected sources that currently exist and should be sufficient until new scholarship and archival materials become available.

Additional support for this work came from *Leatherneck* magazine, *Marine Corps Gazette*, and the *Marine Corps Times*, all of which gave permission to reprint their articles. Their cooperation made this anthology possible. This printing represents as closely as possible the original works as they were published at the time, with minor alterations to the text based on current standards for style, grammar, punctuation, and spelling.

C.P. Neimeyer

Dr. Charles P. Neimeyer
Director of Marine Corps History

Preface

In late 2009, President Barack H. Obama determined that the situation in Afghanistan required a surge of troop reinforcements. For the U.S. Marine Corps, this meant that the Marine involvement in that theater of the Global War on Terrorism would continue to intensify. A Marine expeditionary brigade was deployed to Afghanistan in 2009, where the troop surge and increased Marine presence led to the Marines taking control of security operations for Helmand and Nimroz Provinces in 2010. The Corps' insistence on autonomy within its provinces led to the nickname "Marineistan."

The Marine Corps' initially sporadic participation in Operation Enduring Freedom (OEF) was covered in an earlier volume of this series, *U.S. Marines in Afghanistan, 2001–2009: Anthology and Annotated Bibliography*. This new volume in the series covers Marine operations in Afghanistan from the surge in 2010 to the end of the drawdown in 2014. During the majority of the period covered in this volume, two Marines were the senior American commanders in Afghanistan. General John R. Allen commanded the International Security Assistance Force (ISAF) and U.S. Forces Afghanistan from 18 July 2011 to 10 February 2013. He was succeeded by General Joseph F. Dunford Jr. On 26 August 2014, Dunford was followed by U.S. Army General John F. Campbell.

The Marine Corps approached its increased role in Afghanistan with enthusiasm, employing the Corps' traditional expertise in counterinsurgency (COIN) warfare acquired through the Banana Wars and the Vietnam War as well as lessons learned more recently in Iraq during the al-Anbar Awakening. Along with this focus on the application of COIN doctrine, the Marine Corps fielded the Bell Boeing MV-22 Osprey tiltrotor military aircraft, fully replacing the venerable Boeing Vertol CH-46 Sea Knight helicopters that had been the Corps' primary medium-lift airframe for troops and cargo. Additionally, the recently created U.S. Marine Corps Forces Special Operations Command (MARSOC) fully participated in U.S. Special Operations Command's Afghanistan missions during this period.

In 2010, Marines launched Operation Moshtarak in Helmand Province in an effort to clear the Taliban out of the central part of the province. Combat operations intensified during this period, especially in the Marjah District of Helmand Province, as the Marine Corps pressured the Taliban. At the same time, Marines aided in the intensified training that prepared Afghanistan's military and police forces to take over security of their nation.

Throughout 2011 and 2012, Marine units rotated into Afghanistan and continued to conduct raids and patrols throughout the Marineistan provinces, suppressing the poppy harvest and eliminating Taliban caches and sanctuaries. However, in late 2012, the Taliban launched a successful raid on Camp Bastion, an airfield and logistics base northwest of Lashkar Gah, Helmand Province, destroying six McDonnell Douglas AV-8B Harrier IIs and badly damaging two others from Marine Attack Squadron 211 (VMA-211). Marine aviation personnel fought back as infantry, a role Marines of this squadron last performed on Wake Island during World War II. In the four-hour firefight, all 15 of the attacking Taliban were killed or captured, but Lieutenant Colonel Christopher K. Raible and Sergeant Bradley W. Atwell were killed.

Marines continued COIN and training operations throughout 2013 and 2014, turning over responsibility for security operations to Afghani forces district by district. In May 2014, President Obama declared U.S. combat operations in Afghanistan would end in December of that same year. In October 2014, Marines handed Camp Leatherneck in Helmand Province over to Afghan forces, and in December, the North Atlantic Treaty Organization (NATO) and the United States "ended" combat operations in Afghanistan.

Unfortunately, this act proved premature. Since 2014, the insurgency in Afghanistan has increased, and the U.S. Army has redeployed major units there. Marines remained in Afghanistan serving in joint operations billets in the training mission. The war in Afghanistan has yet to reach a clean conclusion.

This book could not have been published without the professional efforts of the History Division staff. The editors would like to thank Director of Marine Corps History Dr. Charles P. Neimeyer, Deputy Director Paul J. Weber, and former Chief Historian Charles D. Melson. Colleagues Dr. Breanne Robertson, Douglas E. Nash, and Annette D. Amerman provided unflagging professional advice and support. The oral history transcripts included in this anthology were provided by Dr. Fred H. Allison, head of the History Division's Oral History section. Our Editing and Design Branch, led by Angela J. Anderson, was instrumental in transforming the manuscript into a finished product, editing the manuscript, and overseeing the production process.

This anthology is organized into six parts: one section for each year and a final section devoted to a broader overview of Marine participation in the Afghanistan conflict. This work is not meant to be an authoritative history but rather a selected record of Marine contributions to the Afghan war effort as captured by the media and other sources. It is intended to be used as a starting point for the general public and academic researchers.

Paul Westermeyer
Marine Corps History Division

Map 1. Afghanistan provinces, 2008

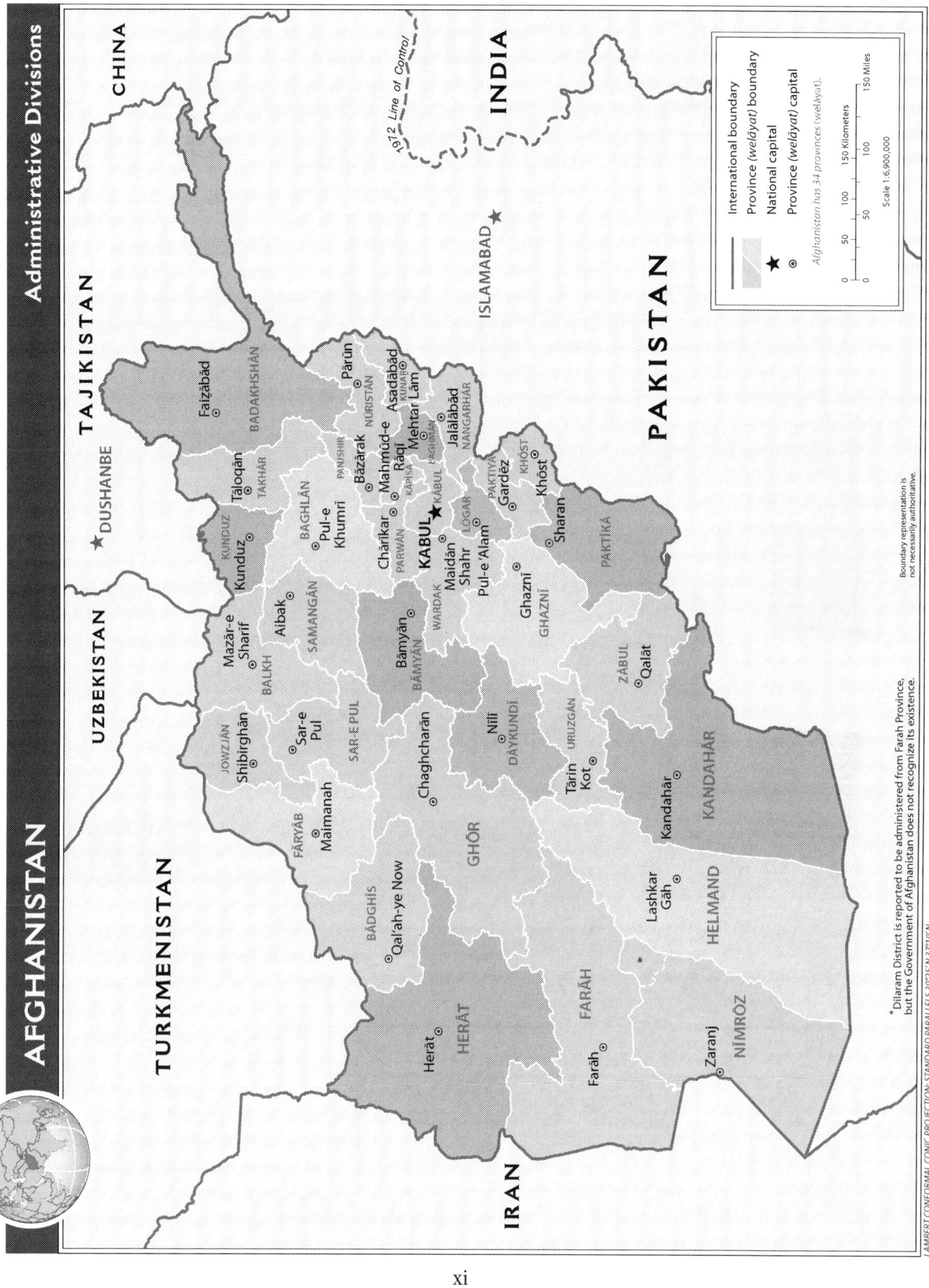

Map 2: Topographic map of Afghanistan

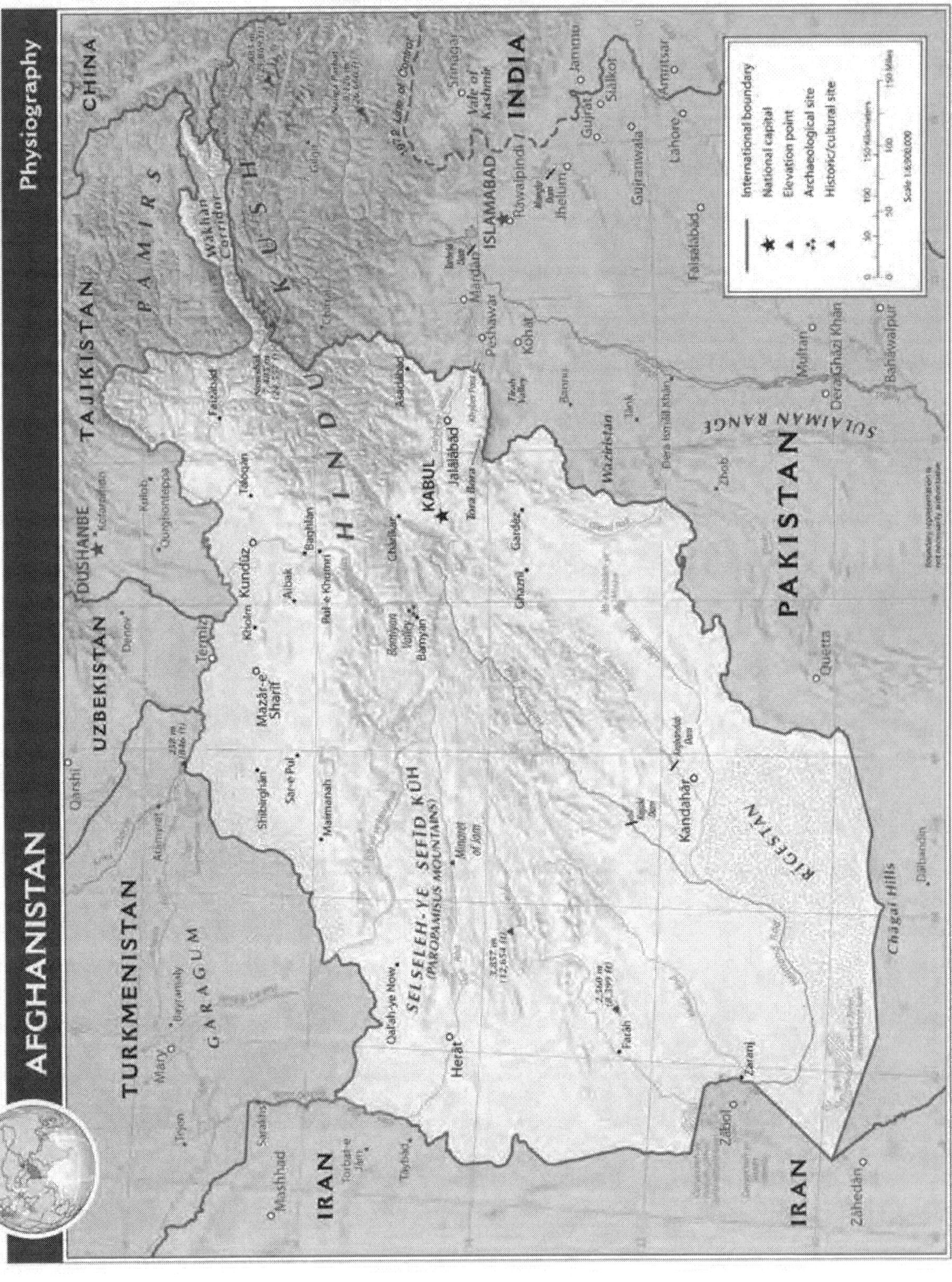

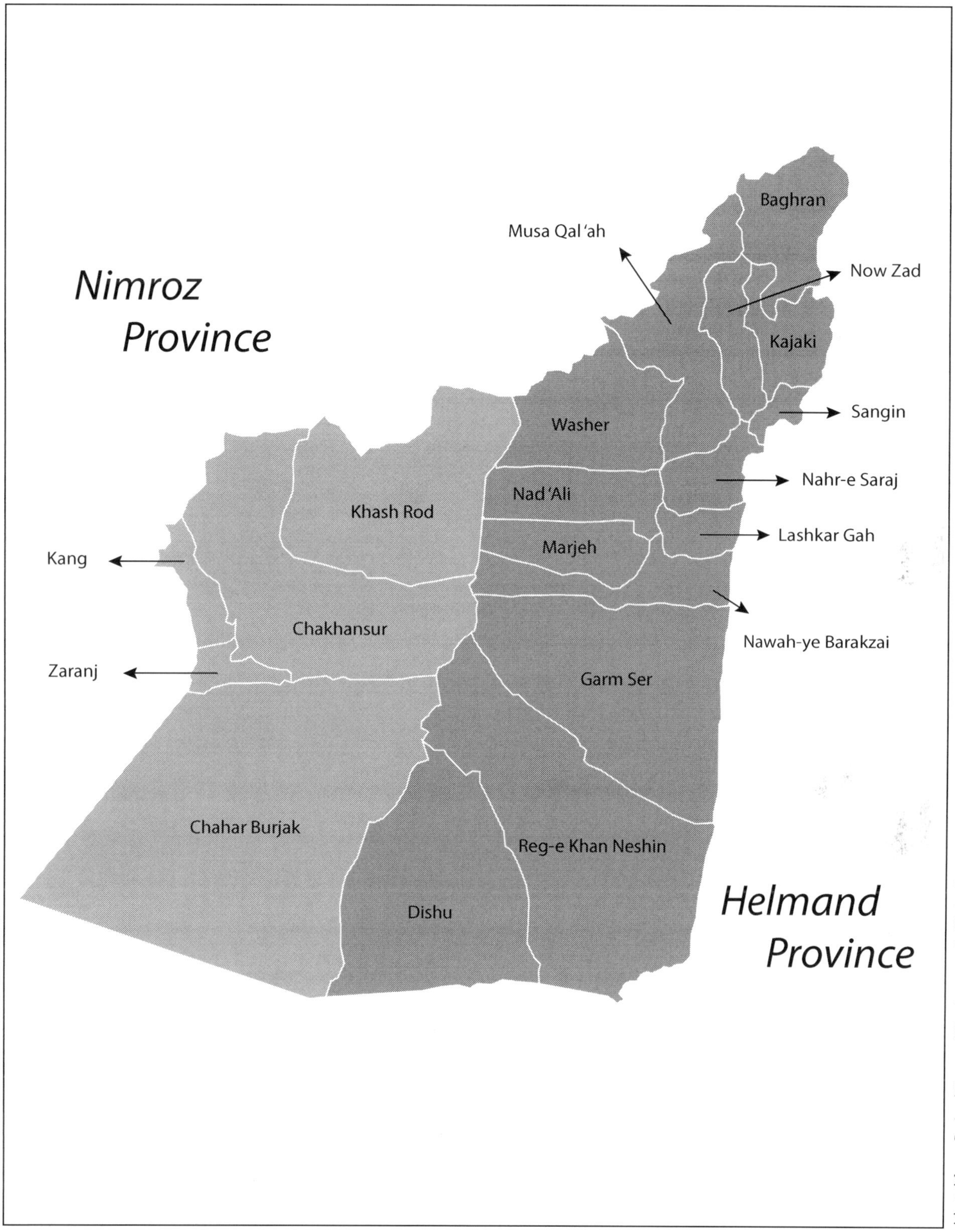

Adapted from a Regional Command-Southwest map by Marine Corps History Division

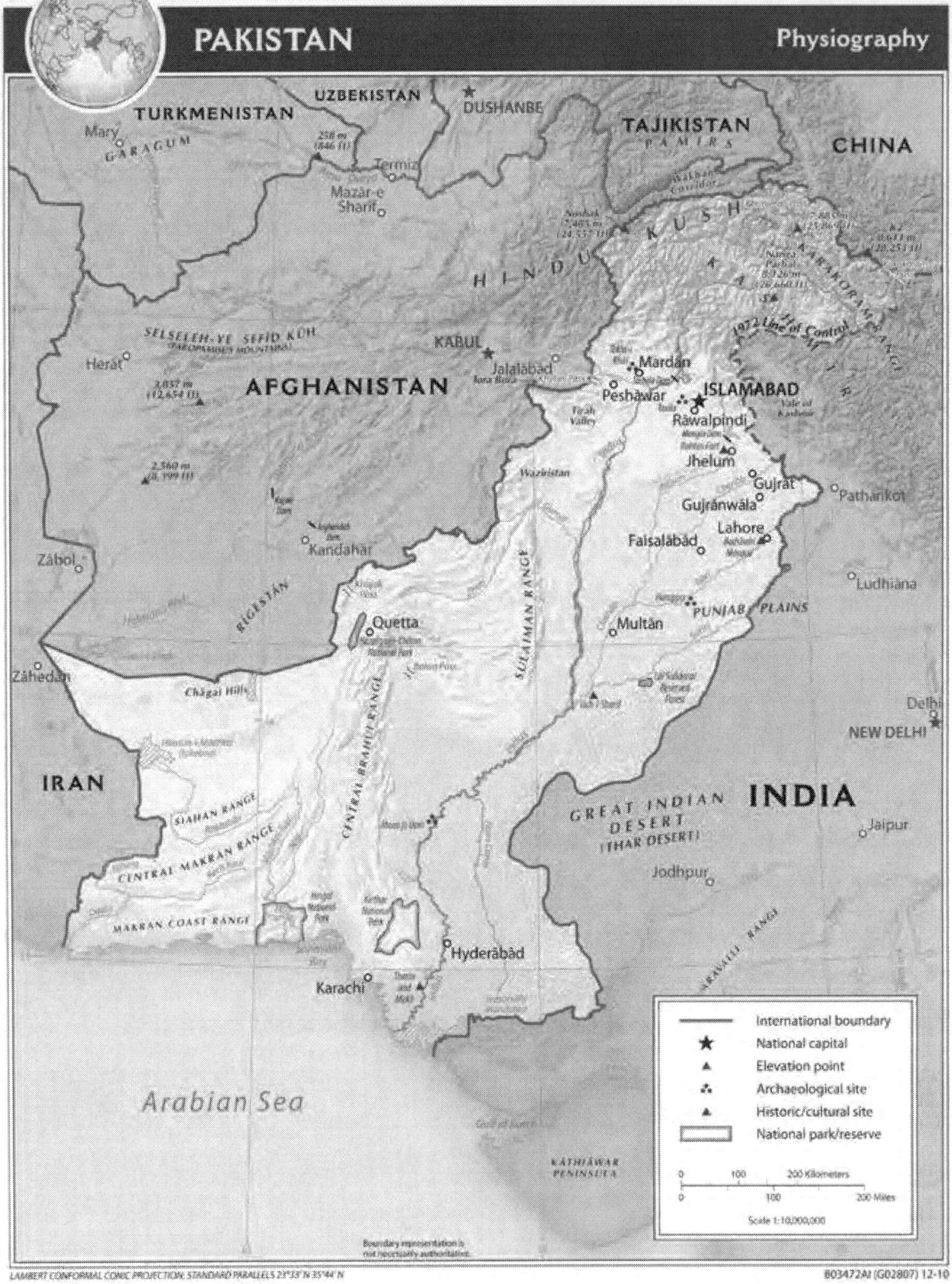

Map 4: Topographic map of Pakistan

Part I
Marineistan, 2010

Gear for Afghan Trip in the Garden by Michael D. Fay.
Marine Corps Art Collection

In 2009, President Barack H. Obama announced a significant increase in American troops for the war in Afghanistan. For the U.S. Marine Corps, this meant intensified involvement in that theater of the Global War on Terrorism. A Marine expeditionary brigade was deployed to Afghanistan in 2009. The troop surge and increased Marine presence led to the Marines taking control of security operations for Helmand and Nimroz Provinces in 2010. Marine insistence on autonomy within its provinces led to the nickname "Marineistan."

In 2010, Marines, fighting alongside British and Afghan forces, launched Operation Moshtarak in Helmand Province in an effort to clear the Taliban out of the central part of the province. Combat operations intensified during this period, especially in the Marjah District of Helmand Province. At the same time, Marines aided in the intensified training that prepared Afghanistan military and police forces to take over security of their nation. In April, I Marine Expeditionary Force (Forward) replaced the 2d Marine Expeditionary Brigade as Regional Command-Southwest.

In September, Afghanistan held parliamentary elections for the Wolesi Jirga (House of the People), but the outcome was clouded by Taliban opposition and internal Afghan government dissension. In November, the North Atlantic Treaty Organization (NATO) announced its plans to end combat operations in Afghanistan in 2014. On 21 November, Lance Corporal William Kyle Carpenter saved a fellow Marine in combat in Helmand Province's Marjah District, an action for which Carpenter was later awarded the Medal of Honor.

An MV-22 Osprey with VMM-261, 3d MAW (Fwd), from Camp Bastion flies over the barren landscape of the Afghanistan desert while performing missions delivering personnel and cargo on 24 April 2010.

Defense Imagery 273191, photo by GySgt Steven Williams

MV-22 Osprey: Proving Itself in Combat

by Corporal Ryan Rholes

Leatherneck, August 2010

In early July, Marine Medium Tiltrotor Squadron 261 (VMM-261) wrapped up its first combat deployment to Afghanistan, with solid proof that the aircraft it flies, the MV-22 Osprey, is ready to replace the Corps' aging fleet of CH-46 Sea Knight helicopters as the go-to medium-lift aircraft for maritime forces.

Getting there has been a long and often controversial progression.

The U.S. government first realized the need for an aircraft capable of vertical takeoff and fast flying in the early 1980s and developed the Joint-Service Vertical Take-Off/Landing Aircraft Experimental [(JVX) aircraft] program in 1981. Bell Helicopter partnered with Boeing Vertol and submitted a plan for an enlarged version of a Bell prototype. Officials designated the JVX aircraft the V-22 Osprey a few years after the initial proposal.

Pilots completed successful helicopter-style, airplane-style, and sea-trial flights with the Osprey in 1989 and 1990, but lost two prototypes in the following two years, casting doubt over the program. After a year of redesign, the Osprey re-emerged with several new safety features and returned to the sky. The program progressed without incident until another two fatal crashes in 2000 that again grounded the revolutionary aircraft. The Corps'

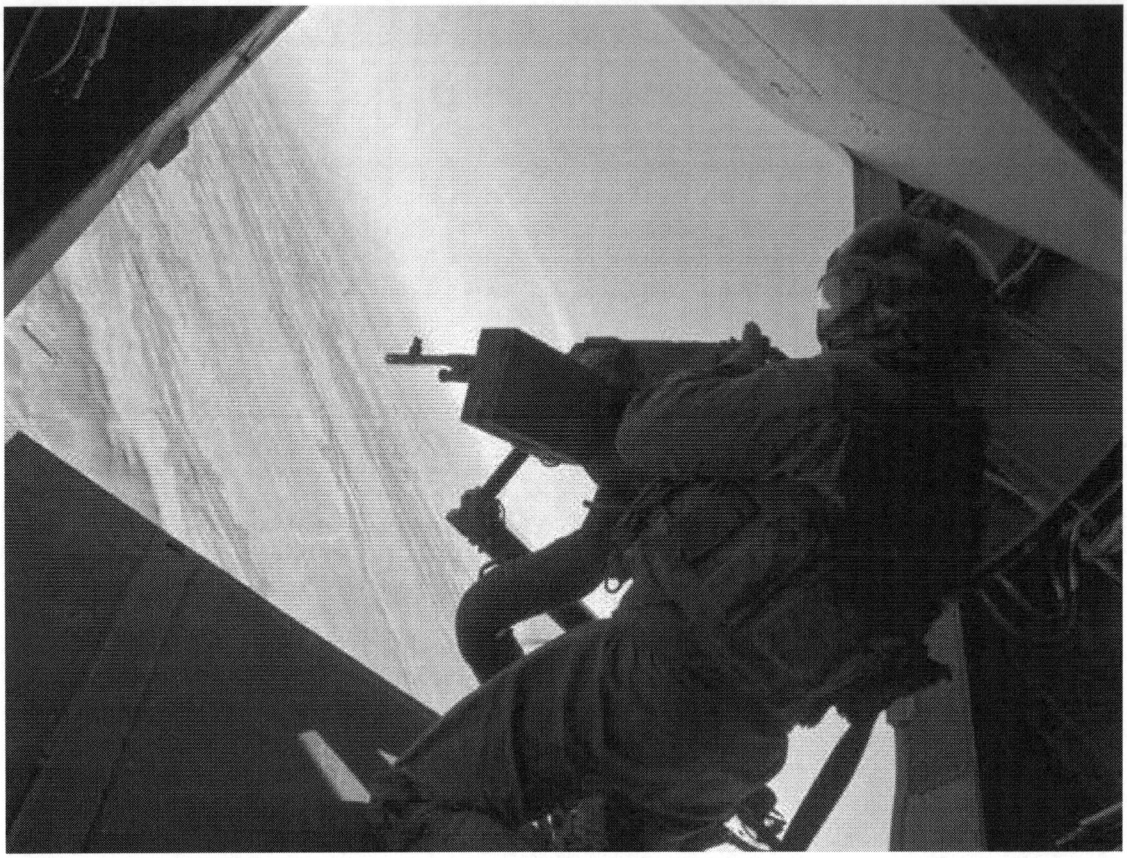

Defense Imagery 273205, photo by GySgt Steven Williams

SSgt John Godwin from Loxfey, AL, serves as a flight equipment technician and aerial observer with VMM-261, where he mans an M240G machine gun on the tail of an MV-22 Osprey while flying missions in Afghanistan.

desire for the Osprey proved resilient and the "bird" underwent heavy improvements, eventually receiving approval from the Pentagon in 2005.

VMM-263 deployed with the aircraft for seven months to al-Asad Air Base, Iraq, in 2007, where it flew more than 2,500 successful missions in al-Anbar Province. Although there was skepticism that the Osprey's success in Iraq would translate to success in Afghanistan, the aircraft would have a chance to prove itself sooner than most expected.

In 2008, most of the Marines serving with VMM-261 had never actually seen an Osprey, according to Lieutenant Colonel Anthony Bianca, the squadron's first commanding officer. In 2009, the squadron taxied an Osprey down its runway at Marine Corps Air Station New River, North Carolina, for the first time, and just a few months later, the squadron was on its way to war.

On 6 November 2009, under the command of Lieutenant Colonel Bianca, several VMM-261 Ospreys flew nearly 600 miles in groups of three from the 22d Marine Expeditionary Unit [22d MEU] onboard the amphibious assault ship USS *Bataan* (LHD 5) in the Indian Ocean to the airfield at Camp Bastion, Afghanistan.

Since arriving, the Osprey has filled a myriad of roles. From troop transport and routine cargo drops, to participating in clandestine special

Defense Imagery 273208,
photo by GySgt Steven Williams

Marines board an MV-22 Osprey on 24 April. This is one of several groups of personnel whom VMM-261 would transport on this day, as the Osprey continues to be a valuable asset for "shrinking the battlespace" by flying faster and farther than any assault support aircraft in the region.

operations drops, the squadron and its aircraft missed no opportunity to prove the aircraft's versatility.

The way an aircraft is used depends on the theater and the needs of the theater commander, said Lieutenant Colonel Ivan G. Thomas, the current [commanding officer] (CO) of VMM-261. "We have been used, very effectively, in a general support role for a lot of the long-range missions that require coming in and out of landing zones versus runways. We have been to Bagram, the eastern border with Pakistan, the western border with Iran and the southern border with Pakistan."

Thomas is an experienced pilot who has flown the Sea Knight and the Osprey in operational environments. He's logged missions flying [Boeing Vertol] CH-46E "Phrogs" in Kosovo, Africa, Albania, and Iraq. Although a diehard [*sic*] fan of the older aircraft, he readily admits that its replacement is bringing some very useful assets to the fight, especially its increased range.

That increased range is due in part to the aircraft's ability to fly higher than the CH-46. The Osprey's increased altitude, according to Thomas, flying at an average height of 10,000 feet compared to the Sea Knight's average 1,000 feet, "helps reduce fuel burn rate and allows pilots an opportunity to fly above bad weather and dust storms."

It also keeps the aircraft outside the range of small-arms fire and rocket-propelled grenades [RPGs]. And although the Osprey usually soars around 10,000 feet, it can reach altitudes of up to 25,000 feet.

However, the Osprey's ability to level off in the clouds is not the only leg up it has on the CH-46. The aircraft also is capable of hauling heavier loads for longer distances, which means more armed Marines or more much-needed cargo delivered farther away or to higher mountain posts. "The Osprey can fly twice as far, twice as fast and with three times the payload," said Thomas.

And even those stats may now be outdated. Since arriving in Afghanistan, the MV-22 Osprey has received a software upgrade and changes to operational guidelines that have seriously increased its capabilities. The aircraft's new software boosts its already impressive speed from 245 knots to between 270 and 275 knots, or more than 300 miles per hour, while flying straight and level. The Naval Air Training and Operating Procedures Standardization program now allows Osprey pilots to take off at 30 degrees of angle, which provides a 50-percent increase over its previous 20-degree cap.

Although these changes allow the Osprey to

Defense Imagery 273209, photo by GySgt Steven Williams

With blades turning, Marines load cargo on a VMM-261 MV-22 Osprey in Afghanistan on 24 April, where the Osprey is proving its worth in battle.

get off the ground faster and to cut down on travel time between destinations, the increased capabilities have not caused any changes in the squadron's doctrine.

"These are great improvements and great capabilities, but it has not dictated that we change the way we fly our aircraft. This deployment has really validated the training [tiltrotor] squadrons receive at Mojave Viper [Marine Corps Air Ground Combat Center, Twentynine Palms, California]," said Thomas. "We are definitely going to concentrate on and reinforce our techniques, tactics and procedures before our next deployment because they have proven effective."

The squadron has flown about 30 named missions since arriving in Afghanistan, in addition to its daily grind of fulfilling assault support requests, all without any serious mishaps. Operating at such a high capacity and in such an austere, harsh environment also has given the squadron ample time to develop and adapt maintenance routines. The increased heat, high winds, and fine dust of Afghanistan cause increased wear on all types of aircraft. The Osprey, which has hundreds of miles of wire running through it, has experienced a few unique problems there.

"The MV-22 is fly-by-wire . . . it's a computer," said Thomas. "Rather than use a push-pull rod that might be hydraulically actuated, you are feeding information to a computer that tells flight controls where to go, which means there is a lot of communication going through a lot of wires. So with the MV-22, there are a lot of wiring issues."

VMM-261's avionics technicians have, in the words of their commander, become "experts," with the aircraft's wiring system. They can quickly "shoot" wires, which means they can check wires from start to finish for the tiniest breaks and rubs that could potentially cause future issues.

However, the avionics Marines are not the only Marines putting in overtime to keep the aircraft up. The squadron performs an average of 7,500 maintenance man-hours each month on its fleet of aircraft. It takes the full maintenance team to keep this squadron at a high rate of readiness.

"The real heroes are the Marines who work on the aircraft," said Thomas. "They are the ones who figure out how to adapt the maintenance and to ensure the aircraft are ready to go."

Keeping this aircraft "ready to go" is not as easy as it sounds because pilots, even though they fly these aircraft well, nonetheless fly them hard.

"One thing I make sure I tell people is that we are not babying [the Osprey] at all," said Brigadier General Andrew W. O'Donnell Jr., the 3d Marine Aircraft Wing (Forward) commanding general who is now in charge of all aviation assets in Regional Command (Southwest) [RC-Southwest] in southern Afghanistan. "That aircraft is landing in the dustiest zones, it flies at night, it flies in low light, and it does everything everyone else is doing, if not more, because of its capability."

Although the Afghan environment places increased strain on the aircraft and causes accelerated engine wear, the problems are not exclusive to the Osprey. "The environment is extremely tough on all the gear," said Brigadier General O'Donnell, "whether you're talking about a generator, an MRAP [mine resistant, ambush protected armored vehicle] or even tents. We have these storms come in, constant extreme temperatures and the 'moon dust' is taking its toll on everything."

However, the Ospreys and their crews have more than increased maintenance requirements to worry about while operating in such a hostile environment.

The Marines of VMM-261 often have found themselves on the front lines, dropping North Atlantic Treaty Organization forces into hostile zones or flying holding patterns above a fray waiting to swoop in with quick reaction forces.

"Everywhere we go where the threat has dictated, we have had [tactical air] escorts," said Lieutenant Colonel Thomas. "Whether it was [McDonnell Douglas] Harriers or [McDonnell Douglas F/A 18] Hornets [attack aircraft], [Bell UH-1 Iroquois] Hueys or [Bell AH-1] Cobras [helicopter gunships], those guys are dedicated to observing objective areas prior to our arrival, and will clear the zone and provide security for us to come in."

Afghanistan has been a learning curve for the crews and a test for the airframes of the still young aircraft. However, it is undeniable that the Osprey is performing on par with the other aircraft on the flight line. Although it had a rocky beginning, the Osprey has settled into a groove and is performing at an exceptional level in one of the most rugged and challenging regions in the world for tactical aircraft.

According to Thomas, the Osprey has moved beyond being an aircraft oddity in the skies over Afghanistan or for that matter anywhere in the world and has taken its place as the Corps' premier medium-lift aircraft.

Note

Leatherneck 93, no. 8 (August 2010): 42–45. Reprinted with permission. Copyright *Leatherneck*.

About the Author

Corporal Ryan Rholes was assigned as a Marine combat correspondent with 3d Marine Aircraft Wing at Camp Leatherneck, Afghanistan.

Sgt Nester R. Hernandez, a field artilleryman, is one of India Battery's provisional mortarmen.

Photo by Cpl Daniel Blatter

Marine Artillery in Afghanistan: 1/11-Every Marine an Infantryman and a Cannoneer

by Andrew Lubin

Leatherneck, March 2011

"We did it all," said Lieutenant Colonel Adolfo Garcia Jr. "The Marines provided 24/7 fire support and conducted full-spectrum COIN [counterinsurgency] operations. We ran convoys, trained ANPs [Afghan National Police] and partnered with the Kajaki District governor."

Garcia commanded 1st Battalion, 11th Marine Regiment (Reinforced), on its 10 May–1 December 2010 deployment to Regional Command Southwest. His artillerymen had perhaps the most robust artillery deployment since 1st Battalion, 10th Marines, fought at an-Nasiriyah in 2003. Garcia's was a composite artillery battalion with 18 M777 155mm howitzers and six High Mobility Artillery Rocket Systems (HIMARS) supporting two regimental combat teams throughout RC-Southwest (Helmand and Nimroz Provinces).

Garcia split his batteries into platoons, placing the powerful howitzers at Firebases Saenz, Zeebrugge, Fiddler's Green, Edinburgh, Pico, and Payne. In all, until Firebase Pico was closed, there was Marine artillery on call at all times throughout Afghanistan's Helmand River Valley. But closing

Photo courtesy of Capt Richard Robinson

Firebase Saenz: Bravo Battery staff, from the left, 1stSgt James R. Telley, 1stLt Benjamin E. Gardner, SSgt Jordan McMahon, Capt Richard M. Robinson (kneeling), GySgt John Clark (kneeling), 1stLt Mark R. Kraus, and SSgt Brian J. Lindsay.

Pico did not reduce available firepower; those guns and Marines were converted to a mobile battery and initially sent to support 3d Battalion, 1st Marines, in the Garmsir District before being employed in other locations.

A high point during the deployment was their assignment as an integral part of Task Force Highlander for Operation Steel Dawn II, joining 3d Marine Aircraft Wing (Forward) and 1st Light Armored Reconnaissance Battalion. Garcia task-organized a HIMARS platoon from S Battery, 5th Battalion, 11th Marines, and a battery-minus from Lima Battery, 3d Battalion 12th Marines, to conduct "shoot and scoots" around the southern city of Barham Chah [*sic*], which had been identified as a critical logistics hub through which the insurgents were receiving and distributing munitions throughout the country. Large amounts of Taliban supplies were destroyed, for as Garcia said, "We shot opportunistically through the length [of] the mission."

Bravo Battery

The "Beastmasters" of Bravo Battery split the battery and brought three M777 howitzers to support the Marine infantry battalions clearing the Sangin River Valley, while leaving the remainder of their tubes at Firebase Fiddler's Green in order to support 1st Battalion, 6th Marines, and 3d Battalion, 6th Marines, in Marjah.

Bravo's effort clearly demonstrated that Marine artillerymen today do more than conduct fire missions. From Fiddler's Green, the battery conducted convoy operations and route security, civil-military operations and COIN operations with the locals. They did this while supporting their Marine infantry brothers in 3d Battalion, 7th Marines, and later 3d Battalion, 5th Marines, in Sangin with pinpoint accuracy.

"We also supported the Georgian [Republic of Georgia] infantry who fought from the Sangin District Center," said Bravo Battery's commander, Captain Richard Robinson. "They [Georgian infantry] were well trained. They understand the COIN fight and knew what was the correct round for the correct mission."

Initially, the Beastmasters shot GPS-guided [Raytheon] Excalibur [precision-guided] rounds. With an accuracy of less than 10 meters out to 24 miles, the rounds kept collateral damage to a minimum. But in spite of the diverse and demanding nonartillery-type missions, Robinson also is very proud of the accuracy the Marines achieved when firing conventional high explosive [HE] rounds. "We had no short rounds—no missed targets. We shot every mission properly regardless of the conditions and the situation.

Photo courtesy of Capt Richard Stinnett

Sgt Jose Saenz shown while on patrol. He was killed in action on 9 August 2010.

"We had some exceptional Marines," Robinson continued. "Corporal Paul Morales and Lance Corporal Andrew Parks each received a battlefield promotion, but frankly most of the Marines went 'above and beyond' every day." Sadly, Sergeant Jose L. Saenz III, a Marine in India Battery, was included in the above and beyond [category]. A vehicle mechanic by military occupational specialty (MOS), he was acting as a provisional infantry squad leader when killed on a mission. The Marines renamed a firebase in his honor.

India Battery

The expression "every Marine a rifleman" was never so evident as with India Battery. Relieving a contingent of British soldiers (C Company, 3d Battalion, The Rifles) at Firebase Zeebrugge, the India leathernecks were tasked with providing security for the five-mile area surrounding the Kajaki Dam. Sitting 35 kilometers to the east of Musa Qala, the American-built dam (1953) has two vital functions: to supply electricity and to provide the water to irrigate some 650,000 acres of Helmand Province farmland.

Unlike their fellow cannoneers at Firebases Fiddler's Green, Saenz and Edinburgh, the India Btry [Battery] "Redlegs" were not supporting any Marine unit other than themselves.* As the battery commanding officer, Captain Richard Stinnett, explained, "Instead of firing to support

* *Redlegs* is an historic term referring to the red stripe on infantrymen's pants.

Adapted from a *Leatherneck* map by Marine Corps History Division

1. Forward Operating Base (FOB) Delaram-2
2. Camp Leatherneck
3. Camp Bastion
4. Firebase Edinburgh
5. FOB Musa Qala
6. Firebase Zeebrugge
7. Sangin
8. Firebase Saenz
9. FOB Marjah
10. Firebase Fiddler's Green
11. FOB Delhi
12. Firebase Pico
13. Firebase Payne

Photo by Cpl Daniel Blatter

Cpl Jacob B. Hoag, India Battery, mans the special application scoped rifle (SASR).

an infantry battalion, there was just us, operating as provisional infantry. There was a South African contractor working for USAID [U.S. Agency for International Development] handling security for the actual dam, and our mission was to clear and control the area surrounding it."

Firebase Zeebrugge had been a British base, built on the grounds of the prior Afghan king's summer retreat. The permanent buildings were helpful. Stinnett commanded a full-strength battery with multiple attachments, including explosive ordnance disposal technicians, mechanics, joint tactical air controllers, electricians, and even two cooks. There were 170 Marines at Zeebrugge. Due to the improvised explosive device (IED) threat, the base could be resupplied only by air. Truck convoys were too dangerous.

"The Taliban had driven the locals out of 12 of 14 local villages but let them return during the day to farm. So we and our ANPs patrolled two to three times daily and began to build a rapport with them," Stinnett explained. As in Nawa, Garmsir, and other villages the length of the Helmand River Valley, the constant Marine presence was effective, and the villagers began to return to their villages and provide intelligence regarding IEDs.

"We were all infantrymen. Our cooks manned the .50-cal. [machine guns] and did overwatch. A mechanic, Corporal Jacob Hoag, was our SASR [special application scoped rifle] gunner, and my

Photo by Cpl Daniel Blatter

Cpl Matthew Lucas, an India Battery cannoneer, serves as a provisional infantryman helping patrol the Helmand River near Kajaki Dam.

Photo by Cpl Daniel Blatter

LCpl Bryant Wahlen, a designated marksman in India Battery, sights in on enemy activity across the Helmand River on 28 September 2010, while patrolling Kajaki Dam.

FDC [fire direction controlman], Corporal Kenneth Saeger, handled the [FGM-148] Javelin [antitank missile]," said Stinnett.

His six provisional infantry squad leaders were led by cannoneers (MOS 0811), radio operators, and a motor transport technician. "We also had two mortar tubes manned by the 0844s, the FDCs. Lance Corporal Jonathan Mera and Corporal Adam Skidmore stood out amongst an excellent FDC team," he said.

With the regimental "gunner," Chief Warrant Officer 4 Thomas R. Beecher, coming out to provide training, they learned quickly, firing some 900 HE and illumination rounds from their M777 howitzers, in addition to the Excalibur rounds.

The daily patrols put India Battery in regular contact with the Taliban, starting the first day when the Brits took them out and became embroiled in a roiling firefight. "We had 162 engagements, ranging from long-distance machine-gun fights in the rolling desert terrain to the north, to being attacked from tree lines and cornfields in preplanned ambushes in the southern green zone. Every IED strike was the beginning of a firefight. We were often being fired on from three positions and unable to bring in CASEVAC [casualty evacuation helicopter]," said Stinnett.

Fighting as infantry took a toll. The India Battery leathernecks suffered 27 wounded and four killed—heavy losses for an artillery battery.

"We had good ANPs. They fought hard, they fought daily. It helped that they all came from this area. The police chief had not seen his home in the six years since the Taliban came; he was motivated to push them out. It looked, by the time we left, that we were making a difference," Stinnett added.

From shooting fire missions with their howitzers, to conducting COIN operations, to supporting their own newly tasked provisional infantrymen with their newly tasked provisional mortarmen, 1st Battalion, 11th Marines, proved that Marines can accomplish any mission.

Note

Leatherneck 94, no. 3 (March 2011): 38–41. Reprinted with permission. Copyright *Leatherneck*.

About the Author

A frequent contributor for *Leatherneck*, Andrew Lubin embedded with Marines in Afghanistan during late May–June 2010 and remains in contact with deployed and deploying leathernecks. His updates from Afghanistan continue to be available on the *Leatherneck* website. Additionally, he is the author of the award-winning book *Charlie Battery: A Marine Artillery Battery Unit in Iraq* (2004), as well as *Keep Moving or Die: Task Force Tarawa at An-Nasiriyah* (2010).

Task Force Highlander's forward command center on the morning of its raid against Bahram Chah on 30 October 2010 was situated on the northern outskirts of the main residential area at the base of a hill in Afghanistan bearing an image of the Taliban flag.

Defense Imagery 101030-M-1263P-011

Operation Steel Dawn II: The Raid on Bahram Chah

by Colonel Nathan S. Lowrey, USMCR (Ret)

Marine Corps History, Summer 2015

Introduction

While speaking at a May 2003 news conference in Kabul, Secretary of Defense Donald H. Rumsfeld declared optimistically that Coalition operations in Afghanistan had "clearly . . . moved from major combat activity to a period of stability and stabilization and reconstruction activities."[1] Benign neglect over the next three years, however, allowed the Taliban to regroup, to adopt new tactics, and to expand their insurgency to the south and east.[2] A British contingent of the International Security Assistance Force (ISAF) assumed command of Regional Command South (RC-South) in 2006, but encountered significant resistance when it attempted to challenge enemy activity in the area, particularly in Helmand Province to the southwest, where the enemy tended to operate with near impunity.[3]

Around the same time—as hostilities in Iraq began to decline—Commandant of the Marine Corps James T. Conway sought a more active role for his Service in Afghanistan.[4] The number of Marine expeditionary forces in southwestern Afghanistan increased, with the type of deployed units growing steadily from a regimental to a brigade and finally to a corps-size presence. Two months after the arrival of Major General Richard P. Mills and I Marine Expeditionary Force (I MEF/Task Force Leatherneck) in April 2010, Marines established Regional Command Southwest (RC-Southwest) in Helmand and Nimroz Provinces.[5] In addition to allowing senior leaders to focus on smaller geographic areas, the new command enhanced the Marines' autonomy and allowed them to maintain an offensive posture while working to extend security throughout the region.

Operation Steel Dawn II—a joint raid against an enemy command center and logistics hub in Bahram Chah, Afghanistan, during October 2010—exemplified I MEF's aggressiveness as well as the operational utility of the Marine air-ground task force. Nestled in a remote desert valley among the Chagai Hills, Bahram Chah is situated in the southernmost district in Helmand Province. Its isolation and proximity to the Pakistani border provided the Taliban with a safe haven ideal for pursuing a wide range of illicit activities. In addition to an al-Qaeda training camp and Taliban prison, the area housed facilities for producing improvised explosive devices (IEDs). The Bahram Chah bazaar served as a central location for smuggling fighters, weapons, ammunition, and explosives northward into Afghanistan and for moving refined opium, heroin, and other narcotics southward out of the country.[6] Major General Mills considered it "a dark and evil place" and decided to deny the enemy sanctuary.[7]

Background

Coalition forces were aware of the town's notorious reputation and had been raiding Bahram Chah on a regular basis for six years. Established by the Taliban in the mid-1990s, the town tripled in size between 2002 and 2005, with as many as 1,000 drug traffickers operating in the area on a busy day. Coalition forces first confronted the enemy sanctuary in May 2005 when the recently established Afghan Special Narcotics Force (ASNF) raided the town's bazaar during a three-day operation to eliminate drug laboratories in southern Helmand Province. Although the drug traffickers fled to safety across the Pakistani border before the raid force arrived, ASNF personnel seized 250 kilograms of heroin, 2.5 tons of opium, and 3.5 tons of precursor chemicals used to manufacture heroin.[8]

The Taliban retaliated four months later by ambushing a company-size Afghan National Police convoy as it traveled south through a canyon located five kilometers north of Bahram Chah. The attack began shortly before nightfall, with insurgents firing down from all directions. When the fighting ended seven hours later, six vehicles had been seized or destroyed and 18 police officers had been killed, including the provincial deputy police chief.[9]

Not to be deterred, the ASNF raided Bahram Chah several times in April 2006. During the first operation, the antinarcotics force seized 75 kilograms of opium resin and arrested one suspect. During the second operation, the force arrested four suspects, seized 1,000 kilograms of opium, and destroyed an arsenal of heavy weapons discovered in an underground bunker. A third raid resulted in the death of one drug dealer as well as the seizure of substantial amounts of heroin, morphine, opium, and drug-making chemicals.[10]

In May 2007, a U.S. electronic surveillance unit (Task Force Orange) tracked Mullah Dadullah to Bahram Chah. Released two months earlier during a controversial prisoner exchange, Dadullah was a senior Taliban military commander and trusted advisor to Mullah Omar, the infamous one-eyed Taliban leader. After a small reconnaissance element determined that air strikes alone

Defense Imagery 101029-M-0000A-001

A light armored vehicle from Task Force Highlander (1st LAR) sits outside the Bahram Chah bazaar on 30 October 2010 during Operation Steel Dawn II.

were insufficient to ensure Dadullah's elimination, Afghan special operations forces and 50 commandos from C Squadron, British Special Boat Service (SBS), were inserted by two Royal Air Force Boeing CH-47 Chinook helicopters. Although immediately targeted by rifles, machine guns, and rocket-propelled grenades (RPGs), the assault force maneuvered forward against the walled compound's 20 defenders. After a four-hour firefight, during which the commandos suffered four casualties, the facility was cleared and Dadullah killed.[11]

In October 2008, Afghani, Pakistani, and Arabic militants gathered at Bahram Chah. Coalition aircraft struck the group with precision munitions during a nighttime raid after ground reconnaissance forces positively identified the insurgents. Although an ISAF spokesman said that the target had been a small number of Taliban commanders, an Afghan official from Helmand Province claimed that two vehicles were destroyed and as many as 70 fighters killed.[12]

Bahram Chah may have escaped the Coalition's attention in 2009, but southwestern Afghanistan did not. Arriving that spring as a vanguard of President Barack H. Obama's eventual troop surge, 2d Marine Expeditionary Brigade deployed to Helmand Province and began clearing operations to break a stalemate that had existed between British and Taliban forces for three years. In July during Operation Khanjar (Strike of the Sword), three Marine battalions pushed south from Camp Bastion and occupied key population centers in Garmsir, Nawa-I-Barakzayi, and Khan Neshin.

The 2d Light Armored Reconnaissance Battalion (2d LAR) occupied Khan Neshin and established the Marines' southern flank in the rural Khan Neshin District. Light armored reconnaissance forces continued to patrol the area during successive troop rotations, conducting counterinsurgency (COIN) operations along the lower Helmand River valley and interdiction operations across the southwest desert. Lieutenant Colonel Michael Martin, who commanded 4th LAR in December 2009, explained that while he would have welcomed an outpost at Bahram Chah, the battalion lacked sufficient resources and the town's isolation would have placed his Marines at risk.[13]

Lieutenant Colonel Scott D. Leonard and 1st LAR (Task Force Highlander) assumed tactical responsibility for Reg-e Khan Neshin District, Helmand Province (Area of Operations Mameluke), in May 2010. Task Force Highlander's headquarters was located at Combat Outpost Payne, situated near the district center in Khan Neshin on the northern side of the lower Helmand River "fishhook." Tasked with disrupting enemy operations throughout the region, Leonard assigned three of his maneuver companies to sectors of responsibility within the battalion's 750-square-kilometer area of operations; a fourth company supported I MEF operations in the Kajaki District to the north.

Alpha and Charlie Companies, commanded by Major John R. Bitonti II and Captain Jason T.

Ford, focused on COIN operations conducted along the north and south banks of the lower Helmand River. These distributed operations had dispersed platoons assigned to areas of responsibility. The Marines patrolled actively to engage the populace and secure the region's thoroughfares, a physically demanding and dangerous mission that often became kinetic. Ambushes or strikes from IEDs buried along the dirt roadways were common and a constant concern. Each Marine company also worked diligently to develop the civil infrastructure, local government, and national security forces in their sectors.[14]

Farther south, operating in a more austere and sparsely populated area, Captain Adrian B. Haskamp commanded Company B. Lieutenant Colonel Leonard realigned the battalion's assets to take advantage of the armored vehicle's mobility and directed Bravo Company to focus on interdicting the enemy's clandestine transportation routes, or ratlines, running through southern Helmand Province. During long-range patrols that lasted up to 30 days and were conducted more than 100 kilometers from the closest support facilities, the company split into platoons and sections to search for contraband and to document the region's demography. Assisted by intelligence, reconnaissance, and surveillance (ISR) assets that helped direct their efforts, the Marines detained 21 suspects for questioning, identified two enemy command and control nodes, and seized more than 4,400 kilograms of opium, 1,750 kilograms of ammonium nitrate, and 70 IED components. The Marines' presence severely restricted Taliban movement and forced the enemy to shift its trafficking routes farther west and east into the desert.[15]

During 2010, as Coalition forces expanded their influence over southwestern Helmand Province, Bahram Chah received a greater amount of military attention. On 22 March, a combined patrol operating in Registan District stopped two vehicles transporting 725 kilograms of hashish to the enemy transshipment center.[16] On 2 June, U.S. Air Force aircraft launched a missile and rocket at enemy positions that had fired on Coalition forces operating near Bahram Chah.[17] Then, on 1 July, Afghan special operating forces and British SBS commandos again raided the town's bazaar.[18]

In the latter action, commandos were inserted via CH-47 helicopters near the village of Haji Wakil. They arrived around 0200 hours and, shortly after, encountered enemy resistance. Forward progress remained difficult, and the commandos had to maneuver through an orchard while being fired on from all directions. The commandos did accomplish their mission, but Royal Marine Corporal Seth Stephens was killed and at least one other team member wounded during the five-hour firefight.[19]

Although clear evidence of enemy activity in Bahram Chah was found, the military operation had been costly. The commando's ground assault force was unable to sufficiently support the withdrawal of the raid force, and 1st LAR—despite being designated the regional quick reaction force—was not contacted when the situation began to deteriorate. On 19 September, Coalition aircraft launched two rounds of precision munitions at the town's bazaar, destroying an IED factory and 4,500 pounds of explosives material.[20]

Planning and Preparation

Lieutenant Colonel Leonard and those from Task Force 210 discussed the problems with Bahram Chah that summer, but waited until early October to propose a raid of the enemy's logistics base. The proposal was presented to Brigadier General Joseph L. Osterman and 1st Marine Division (1st MarDiv Forward) staff at Camp Leatherneck.[21] During an operational debrief, Leonard and his

Defense Imagery 140514-M-RD023-096, photo by Cpl Ismael E. Ortega

Marines with Bravo Battery, 1st Battalion, 11th Marines, fire the M777A2 howitzer during Exercise Desert Scimitar 2014 at Marine Corps Air Ground Combat Center Twentynine Palms, CA.

British counterpart described the tactical situation and pitched the concept for near-simultaneous air and ground assaults. The plan called for two companies driving down from the north to screen the bazaar from adjacent residential areas to the west and east, enabling the commandos to fly in and strike the objective.[22] By attacking before the Taliban had an opportunity to withdraw southward into Pakistan for the winter, Coalition forces could delay the production of IEDs and disrupt enemy operations into the summer of 2011.[23]

Around 10 October, 1st MarDiv endorsed the concept, adding slides to illustrate the mission before submitting it to I MEF (Forward) five days later.[24] Major General Mills directed the planning to continue and forwarded the proposal to the ISAF Joint Command (IJC) for approval. In the end, the plan took about two weeks to wend its way up the chain of command. At some point, ISAF briefed the Afghan government, which in turn notified the Pakistani government 24 hours in advance of the raid, enabling the Pakistani Army to withdraw its border position—but this may have also alerted the Taliban.[25]

Most of the deliberate planning and preparations occurred at Combat Outpost Payne.[26] One of the more difficult issues the planners had to address was targeting. Although the first 140 kilometers of the proposed route to Bahram Chah included open desert, the final 20 kilometers involved traveling through one of two narrow mountain passes to reach the objective. Aware of the region's recent operational history, the Marines knew they would have to eliminate the enemy's defensive positions along the pass in advance of the movement. They developed the plan according to doctrine, incorporating shaping fires, suppressive fires, and fires in support of obstacle breaches. They also observed Standard Operating Procedure (SOP) 398, which governed the use of supporting fires in theater. Captain Nicolas S. Rapkoch, 1st LAR's fire support coordinator, spent many evenings in discussion with the battalion lawyer, struggling to address the engagement requirements before seeking ISAF approval through the IJC.[27]

Operation Steel Dawn II would ultimately involve "one of the most robust fires packages ever executed in Afghanistan," including a wide variety of offensive aircraft: Rockwell B-1 Lancer bombers, Fairchild Republic A-10 Thunderbolt II attack aircraft, Lockheed AC-130 Spectre gunships, McDonnell Douglas F/A-18 Hornet fighters, and Bell AH-1 Cobra attack helicopters.[28] While the IJC may have understood the rules of engagement, Captain Rapkoch later complained, the command did not necessarily appreciate their spirit or intent. The IJC chose to deny some pre-assault fires against fighting positions associated with residential compounds near Bahram Chah because the targeting requirement for two corroborating sources of intelligence in habitable areas was not met.[29]

Marine planners understood that dust or thunderstorms could inhibit the use of aircraft, so an artillery presence would be necessary to ensure an all-weather fire support capability.[30] That task fell to Lieutenant Colonel Adolfo Garcia Jr. and 1st Battalion, 11th Marines. The artillerymen were already well known to 1st LAR—the two battalions had worked together during an Enhanced Mojave Viper exercise at Marine Corps Air Ground Combat Center Twentynine Palms, California, shortly before deploying. After arriving in Afghanistan in May 2010, the 1st Battalion, 11th Marines, established its headquarters at Fire Base Fiddler's Green. The headquarters, one rocket battery, and three cannon batteries operated eight firing positions distributed throughout I MEF's area of operations. In addition to supporting Regimental Combat Teams 2 and 7, the 1st Battalion, 11th Marines, had also secured a portion of Route 605 in Marjah and conducted COIN operations in the Kajaki District.[31]

Because Bahram Chah was located outside the range of the artillery pieces positioned at Combat Outpost Payne, artillerymen had to accompany the raid force. To facilitate coordination during the operation, Lieutenant Colonel Garcia formed an artillery group under the command of his operations officer, Major David J. Grabow. This special purpose task organization included the battalion's jump command post; 1st Platoon, L Battery, equipped with three 155mm light howitzers; and 3d Platoon, S Battery, equipped with the High Mobility Artillery Rocket System (HIMARS).[32]

In preparation for the raid, S Battery, under Captain Jeffrey S. Curtis, conducted test fire ranges at Camp Leatherneck on 11 October and then at Combat Outpost Payne on 14 October.[33] The 3d Platoon also performed a route reconnaissance to a proposed assembly area to ensure the desert terrain was suitable for supporting the heavy HIMARS during travel.[34] In the meantime, Captain Bitonti and Alpha Company received a warning order, switched areas of responsibility with Charlie Company, remounted their armored vehicles, and embarked on an intense week of refresher training and maintenance.[35]

On 19 October, organizations participating in the operation moved toward Combat Outpost

Defense Imagery 140518-M-RD023-077,
photo by Cpl Ismael E. Ortega

Marines with Sierra Battery, 5th Battalion, 11th Marines, fire the M142 HIMARS during Exercise Desert Scimitar 2014 at Twentynine Palms.

Defense Imagery 339079,
photo by Cpl Paul D. Zelner

A convoy from CLB-3, 1st Marines Logistics Group (Fwd), heads to Combat Outpost Payne on 21 October 2010 to support Operation Steel Dawn II.

Payne, which served as an initial staging area for Task Force Highlander. The 1st Battalion, 11th Marines' communications platoon, under Captain Ronnie L. Creech, began to actively support Operation Steel Dawn II, providing the equipment and personnel necessary to operate the forward combat operations center.[36] Lieutenant Colonel Leonard issued an operations order to key leaders the following day and, on 22 October, the participants reconvened to conduct a detailed rehearsal of concepts drill. Majors Matthew Miller and Jon A. Custis, the 1st LAR's operations and executive officers, respectively, walked the participants through the operational sequence and corrected problems identified during the group exercise.[37] Lieutenant Colonel Garcia later noted that "The thorough planning and rehearsals conducted prior to the raid ensured the seamless integration of these elements and refined the command and control of the raid force."[38]

Alpha and Bravo Companies departed Combat Outpost Payne the next day, heading in opposite directions to conduct interdiction operations for five days. As part of Lieutenant Colonel Leonard's deception plan, the companies intended that any Taliban groups monitoring the Marine units' activities would interpret the movements as routine patrols.[39] Shaping and reconnaissance operations conducted in support of Steel Dawn II (e.g., Operations Steel Dawn I and Press 16) also began on 23 October.[40] Gunnery Sergeant Aaron J. Abrams, Alpha Company's 2d Platoon sergeant, served as Task Force Highlander's liaison to Task Force 210 during Operation Press 16. The British reconnaissance patrol collected information on the proposed route, identified enemy observation posts covering the mountain passes, and observed activities in and around Bahram Chah.[41]

On 25 October, Major Custis began to push command and support elements south. The intermediary goal was to establish a logistics staging area (LSA) about 60 kilometers from Combat Outpost Payne.[42] Lieutenant Colonel Leonard, who considered Bahram Chah the "nexus of all bad things . . . happening" in Helmand and Kandahar Provinces, spoke candidly to his Marines before departing: "We will go into the Bazaar of Barham [*sic*] Chah and say, 'We'll come here any time we want—you can't stop us. You don't get to operate with impunity.' We will tell them, 'If

Defense Imagery 101021-M-4388Z-004, photo by Cpl Paul D. Zelner

LtCol Scott D. Leonard, commanding officer of 1st LAR and Task Force Highlander, addresses Marines participating in Operation Steel Dawn II at Combat Outpost Payne on 25 October 2010.

you want to come back and rebuild this place, go ahead. We will be back in three months'."[43]

The LSA was situated in a large *wadi* (dry stream bed) in the middle of the desert, a site specifically chosen for its remoteness. Task Force Highlander assembled 178 vehicles and 1,100 personnel into a raid force in the isolated location without being observed. Getting to the LSA was difficult as sand dunes and rough terrain hindered the progress of the heavily laden vehicles.[44] During the weeklong operation, for example, one M88 armored vehicle crew recovered or repaired 53 vehicles.[45]

Upon the artillery group's arrival at the LSA, Major Grabow established a command and control node to support other task force elements during the operation. The communications platoon from 1st Battalion, 11th Marines, provided host services via multiple tactical satellite radio networks. Due to its communications capabilities, the artillery command element also fulfilled the role of camp commandant.[46]

Combat Logistics Battalion 3 Marines provided critical assistance to Task Force Highlander throughout the operation. Motor Transport Bravo Company, under Captain Matthew J. Neely, provided eight pallets of food and 15,000 gallons of water as well as fuel, ammunition, and other supplies via combat logistics patrols and helicopter support teams. Marines from Landing Support Platoon, Headquarters and Service Company, off-loaded supplies at multiple sites,[47] and the 1st Medical Battalion's Shock Trauma Platoon provided a mobile trauma bay.[48]

Marine Wing Support Squadron 373 also

contributed significantly to the operation, constructing a forward arming and refueling point (FARP) near the LSA.[49] A Bell Boeing MV-22B Osprey from Marine Medium Tiltrotor Squadron 365 (VMM-365), commanded by Lieutenant Colonel Craig C. LeFlore, was the first aircraft to land at the FARP on 29 October and provided assault support during the operation.[50] Captain Patrick W. Richardson, a Bell AH-1 Super Cobra pilot with Marine Light Attack Helicopter Squadron 369, also operated from the FARP. Richardson later explained that "We were able to spend 15 minutes in transit, 20–30 minutes to refuel and rearm, and get back to support troops on the ground."[51] The facility also had on-hand casualty evacuation capability.

On 26 October, operating farther south near the objective area, Gunnery Sergeant Abrams and the British reconnaissance patrol conducted a night mortar raid against Bahram Chah. The commandos established two observation posts—situated on high ground running along each side of the proposed route—and fired at enemy positions previously designated as on-call targets. When the rounds began to impact, the enemy fired flares into the night sky and attempted to flee in vehicles. Abrams reported that numerous headlights illuminated a large traffic jam.[52]

Defense Imagery 101029-M-4388Z-003, photo by Sgt Derek Carlson

Supplies are offloaded from a Bell Boeing MV-22B Osprey with VMM-365 at a remote logistics staging area on 29 October 2010.

The next day, the reconnaissance patrol rejoined Task Force Highlander at the LSA and delivered their account to Lieutenant Colonel Leonard. Although Leonard had originally intended to follow the western pass through the mountains, the patrol reported that the circuitous route would take the convoy directly past the city's main residential area. Shortly before departure, the route was changed to the eastern pass.[53]

Movement to Contact

The raid was originally scheduled to begin on the evening of 28 October, but torrential rains resulted in a one-day weather delay.[54] Around 1800 hours the following evening, Task Force Highlander formed a column and headed south to Bahram Chah.[55] The Marines first needed to reach the line of departure on the north side of Chagai Hills—20 kilometers short of the bazaar—and then maneuver through the mountain pass to arrive at a designated release point by 0700 hours the following morning.[56] The time requirement deepened the Marines' sense of urgency; if they failed to arrive on time, Task Force 210 would have to execute its portion of the mission without ground support. Lance Corporal [Adam B.] Ramirez, a vehicle driver from Alpha Company, later recalled that the Marines had traveled light and fast—bringing only water, chow, ammunition, and a one-day pack per man.[57]

Alpha Company held the lead position in the 85-vehicle convoy, and Captain Bitonti divided the unit into two sections for movement: 2d Platoon with attachments from 3d Combat Engineer Battalion, company headquarters, and logistics vehicles followed by 1st and 3d Platoons. The engineering assets included assault vehicles with rocket-propelled mine clearing line charges and a D7 bulldozer and M870 semitrailer for breaching three IED-laden choke points. The battalion's jump command post was next in line, followed by Bravo Company and the logistics train.[58] Major Grabow and the cannon platoon brought up the rear, while Captain Curtis and the HIMARS platoon remained at the LSA.[59] Major General Mills and representatives from several U.S. government agencies also accompanied the raid force.[60]

Working its way south, Alpha Company marked the route and relayed situational information to the remainder of the convoy. Gunnery Sergeant Abrams, who guided the lead platoon, recalled that the sky had been clear when the company began the road march. However, a solitary black cloud followed the Marines, eventually bringing rain, thunder, and lightning. During the approach, the Marine company had expected to strike numerous IEDs, but none were encountered.[61]

Task Force 210 approached the objective area aboard CH-47 helicopters, relaying position reports to Marines as it crossed airborne phase lines. When the British were approximately two minutes from the landing zone, at around 1940 hours, Task Force Highlander initiated the kinetic portion of the operation.[62] A B-1 bomber attacked enemy fighting positions and observation posts along the Marines' axis of advance and near the commandos' ingress point. During two passes, the bomber struck eight preplanned targets with laser-guided 2,000-pound bombs.[63]

As the convoy reached the north side of the Chagai Hills, the logistics train and artillery group pulled off the road and established a functional firing position in only 15 minutes.[64] While Task Force Highlander maneuvered along the eastern corridor toward Bahram Chah, the artillerymen shot illumination rounds over the western pass in an effort to deceive the enemy on which direction the Marines would initiate the attack. When Alpha Company approached the first IED-laden choke point, the Marines spotted Taliban fighters massing along high ground paralleling the canyon. At Captain Rapkoch's request, the cannon platoon began to shoot preplanned preparatory fires against enemy defensive positions, shifting in concert with the task force's advance, effectively suppressing the enemy with at least one sympathetic detonation and allowing the convoy to continue forward. At a second choke point, the cannon platoon again suppressed Taliban forces.[65] In all, the platoon fired 107 high explosive rounds in support of the breaches.[66]

Task Force Highlander also employed close air support to strike the enemy.[67] Lance Corporal Ramirez recalled the surreal experience of riding through the night with bombs exploding near the objective, artillery impacting along the ridgelines, and illumination rounds drifting downward over the western pass.[68] Sergeant Stuart R. Sanford, a vehicle commander in Bravo Company, similarly remembered an "awesome fireworks show" as the "artillery was blowing the hell out of everything."[69]

The journey south was not without difficulties for Coalition forces. At one point, the trailer hauling the bulldozer became stuck and, when Marines tried to pull it free, an axle broke. While leaders contemplated what to do, a lance corporal climbed into the driver's seat, backed the bulldozer off the trailer, and asked "Hey, gunny, which way do I go?" The gunnery sergeant pointed south, and the young Marine moved out at three and a half kilometers per hour.[70] On another occasion, the combat engineer support element led a convoy down the wrong corridor. When 3d Platoon Commander First Lieutenant Michael D. Wright realized what happened, he initiated corrective action and, on Captain Bitonti's orders, assumed maneuver control of the engineer element.[71]

Defense Imagery 101029-M-1842C-002, photo by Sgt Derek Carlson

Artillery illumination and a desert storm light up the night sky over the Chagai Hills as Task Force Highlander maneuvers south through a mountain pass toward Bahram Chah on 29 November 2010.

Progress was swift and Task Force Highlander reached its designated release point, a hilltop near the southern entrance to the mountain pass, around 0430. Having arrived two hours before sunrise and well ahead of schedule, the Marines received word that Task Force 210 was not ready to launch the attack and halted the convoy. As Marines established a hasty blocking position and waited in the dark, the infrared strobe lights on the antenna of the armored vehicles reminded Captain Rapkoch of a traffic jam on U.S. Interstate 95.[72]

Shortly after being inserted about three miles southeast of the bazaar, the British commandos began to receive small-arms fire. In response, the HIMARS platoon fired three GPS-guided rockets into the buildings where the gunfire emanated. Before long, the commandos proceeded to secure the perimeter. They established preliminary positions on each corner of the bazaar and an observation post on a mountain peak northwest of Bahram Chah; they also marked lanes for the armored vehicles to follow. The Marines then surged forward.[73]

Actions on the Objective

The road sloped down beyond the release point, first heading southeast and then cutting sharply to the southwest before opening into a three-kilometer-wide valley.[74] The smaller village of Jumakhan was situated to the east, the larger village of Haji Wakil to the west, and the bazaar near the middle. When First Lieutenant Sean T. Knapp led the engineer platoon forward in the dark to breach the entrance to the objective area, the platoon encountered small-arms, machine-gun, and RPG fire. First Platoon, Alpha Company, under First Lieutenant Jonathan R. Walaski, moved forward to establish a temporary blocking position and to provide cover fire; the company quickly suppressed the enemy. After the rocket-propelled line charges twice misfired, Captain Bitonti pushed forward with the mine rollers. Around this time, the bulldozer emerged from the pass.[75]

The Marines 1st Platoon quickly maneuvered past the engineers, crossed into the objective area, and swung east to establish the assigned blocking position along the southwest side of Jumakhan. Although intelligence had predicted little if any resistance from the village of single-story adobe dwellings, as soon as the Marines pulled into position, they encountered additional small-arms, machine-gun, and RPG fire. The Marines again suppressed the enemy with ease.[76]

With Lieutenant Walaski's platoon shielding the mountain pass exit, the remainder of Task Force Highlander was able to flow unimpeded into the objective area.[77] Alpha Company passed through the British lines while skirting along the north side of the bazaar and then swung south to cordon off the southwest side of Haji Wakil.[78] Master Sergeant Curtis C. Gregory, Alpha Company's operations chief, recalled a relative calm at that point, as flashes from the commandos' infrared lights and weapons sights could be seen atop buildings in the bazaar[79]—good news for the Marines, who had been precluded from firing their weapons until they passed the British lines.[80]

Captain Bitonti established a linear blocking position, facing west toward Bahram Chah. Lieutenant Michael D. Mitchell's 2d platoon was located on the southern flank in a low-lying, vegetated area near the Pakistani border. The rest of the company was situated on higher ground, with headquarters in the middle and Second Lieutenant Wright's 3d Platoon to the north.[81]

It was dark by the time 2d Platoon pulled into position. The enemy could hear, but not see the Marines' armored vehicles. Before long, the Marines spotted enemy fighters maneuvering toward their position, ducking in and out of buildings and crawling through the brush. The insurgents operated individually, as pairs, and occasionally in teams of three or four. Most were armed with assault rifles, although some carried machine guns and RPGs. In one instance, Gunnery Sergeant Abrams's crew killed a pair of fighters who had tried to fire an RPG at the Marines. When the enemy fell, another fighter picked up the RPG and tried to fire, but was also shot. Nearby, Lieutenant Mitchell's crew similarly engaged a fighter armed with an RPG who was maneuvering through vegetation about 75 meters away. The Marines may have killed a suicide bomber—a man wearing a vest appeared to explode as he ran at their vehicle and was shot.[82]

The enemy maneuvered like well-trained and disciplined soldiers, yet they were obviously unfamiliar with the light armored vehicles' capabilities.[83] In addition to being well armed and impervious to small-arms fire, the vehicles had thermal imagery devices that allowed the crews to shoot accurately in the dark at up to 3,000 meters. According to Sergeant Sanford, the thermal imaging devices made night "as bright as day and so detailed you can determine what kind of weapon the enemy

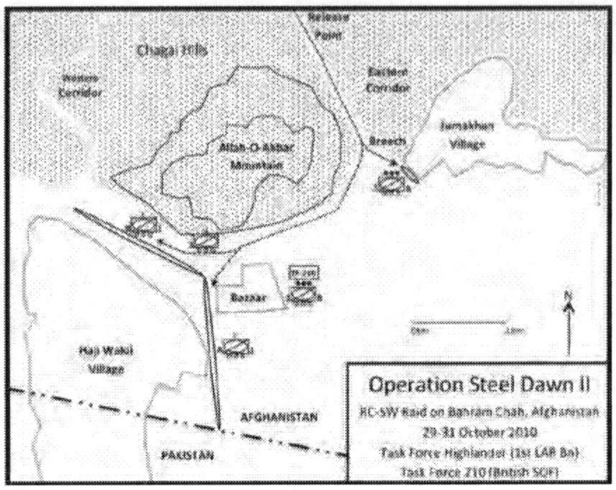

Map by author

is carrying."[84] Gunnery Sergeant Abrams also acknowledged being able to see the enemy maneuvering through buildings and behind walls.[85]

After passing through British lines, Captain Haskamp established a second blocking position with Bravo Company facing south along the north side of Haji Wakil. Second Platoon, under First Lieutenant Gabriel M. Lavine, was positioned on the left flank, with the company executive officer, First Lieutenant Charles L. Hostetler, and Headquarters Platoon in the middle. Third Platoon, under First Lieutenant Christopher M. Phifer, was on the right flank. Each platoon had to carefully observe their sectors of fire, because they were unable to see Alpha Company to the southeast. In addition to cordoning off the residential area, 3d Platoon was also positioned to cover the entrance to the western corridor.[86]

Shortly after Bravo Company passed the bazaar, a vehicle crew spotted and killed three fighters armed with RPGs. Once in position, other members of the company engaged small groups of fighters who were maneuvering among the buildings and through the bush. First Sergeant Jon D. Jerome, unimpressed by the enemy's lack of situational awareness, was dumbfounded that they continued to use paths that the Marines covered with automatic weapons.[87]

Defense Imagery 379620, photo by SSgt Jeremy Ross

A 17 March 2011 photo shows the Bahram Chah bazaar, which facilitated the movement of enemy fighters, weapons, and ammunition into—and drugs out of—Afghanistan.

Defense Imagery 336189

British special operating forces and Marines from Task Force Highlander search a compound at one end of the Bahram Chah bazaar on 30 October 2010.

Meanwhile, Lieutenant Colonel Leonard situated his jump command post on the edge of the objective area, north of the bend in the battalion's L-shaped blocking position, about 800 meters from Haji Wakil.[88] Major Custis established a multiuse point directly behind the objective area, which incorporated the casualty collection, maintenance collection, detainee transfer, and combined-operations coordination points for Task Forces Highlander and 210.[89] Captain Rapkoch took control of all supporting fires and employed AC-130 gunships to disrupt the insurgents' haphazard defenses.[90] According to intelligence reports, little danger of civilian casualties existed because Taliban forces had driven them out [of] the area, and the civilians had not returned before the Marines' arrival.[91] To prevent accusations of firing across the Pakistani border, Marines also used video to capture all weapons fires within the objective area.[92]

Taliban resistance decreased as dawn broke, although isolated individuals and locations were targeted throughout the day. In one instance, for example, Bravo Company Marines observed an enemy spotter using a cell phone to direct 107mm rocket fire. Marines responded with a tube-launched, optically tracked, wireless-guided (TOW) missile launched at the building in which the spotter was hiding.[93]

Task Force 210 cleared the bazaar—which measured approximately 400 meters by 700 meters and contained about 150 shops—in 12 hours. First Lieutenant Buck A. Bradley's 1st Platoon detached from Bravo Company upon entering the objective area and assisted the British by providing cover as the commandos maneuvered forward, suppressing enemy fighting positions and securing high-speed avenues of approach.[94] The commandos moved systematically from west to east and examined each of the bazaar's shops for indications that its occupants were contributing to the insurgency. If the search failed to find such evidence, the shop was left intact. If evidence of illicit activities was discovered (e.g., weapons, munitions, or bomb making or drug processing materials), the material was seized for further exploitation or destroyed on site.[95] During the clearing operation, commandos killed a suicide bomber before he could detonate his explosives.[96]

Captain Bitonti and Master Sergeant Gregory coordinated the medical evacuation of two Afghan soldiers injured while clearing the bazaar. As the two Marines moved toward the bazaar to collect the wounded, a HIMARS rocket struck and destroyed a nearby building, but they were uninjured. One of the Afghan soldiers had been shot in the hip, and the other had twisted his ankle.[97] After transporting the wounded to the multiuse point for treatment and evacuation, the Marines rejoined Alpha Company.[98]

Tensions grew throughout the day as a crowd of unarmed men gathered on the Pakistani side of the border. The men moved about in groups of approximately 10–50 individuals and eventually totaled about 500 people.[99] Gunnery Sergeant Abrams recalled that the men sat on the ground, under trees, on walls, or in compounds. They watched the Marines with casual disregard, as if they were certain of their safety and waiting patiently for the Coalition forces to leave.[100] The prevailing opinion was that the men were Taliban reinforcements, assembled from villages throughout the region, preparing for a counterattack that failed to materialize after the fatal morning defeat.[101]

Later that afternoon, the Marines were able to vent their frustrations over the growing crowd south of the border. A large hill mass north of the objective area overlooked Haji Wakil. On one slope, the enemy had inscribed "Taliban Pass" using white rocks. When A-10 aircraft destroyed the inscription with a 25mm cannon, Captain Bitonti likened the scene to "watching the eraser tool on [Microsoft] Paint."[102] The Marines cheered, but the Taliban were "livid, jumping around and slamming their fists on the ground."[103] Another slope held the image of the Taliban flag with an accompanying inscription; however, a friendly observation post was located nearby, and the site was left intact.

Later that afternoon, insurgent activity escalated, with occasional teams of two enemy fighters emerging from buildings to take potshots at the Marines. Once night fell, scattered engagements of growing intensity occurred across the Marines' frontage.[104] Originally, the combined task force was to remain in the objective area for a second day to clear Haji Wakil, but element leaders reckoned that the most important part of the mission was accomplished and that lingering any longer would tempt fate. The Coalition had suffered two casualties by that point, and the force had gone without sleep for 36 hours. Moreover, Haji Wakil was larger and more complex than the bazaar; the plan for Haji Wakil was not as well developed; and maintaining command and control would be difficult.[105]

At the conclusion of the clearing operation, 1st Platoon, under Lieutenant Bradley, escorted Task Force 210 to an extraction point, located about four kilometers from the objective area.[106] The British were lifted out by helicopter around midnight. After confirmation of the commandos' extraction, Task Force Highlander began its own move from Bahram Chah. Before leaving, however, the Marines destroyed the Taliban flag on the nearby hillside.[107]

Task Force Highlander returned along the same route it had used to enter. The convoy departed by echelon, in reverse order of its arrival, while the cannon platoon covered its withdrawal with smoke rounds.[108] The only thing slowing the Marines' exit was the recovery of abandoned, inoperable armored vehicles from a platoon that had guarded the corridor during the raid.[109]

The convoy reached the LSA around 0500 hours on 31 October. Alpha and Bravo Companies quickly established a southward facing screen in front of the assembly area.[110] While unit leaders participated in an operational debriefing, the Marines attended to their vehicles, weapons, equipment, and personal gear, but also found time to get much needed rest. The following morning, Task Force Highlander began to collapse the LSA.

Operation Steel Dawn II

In late October 2010, Coalition forces from Regional Command Southwest (RC-Southwest) raided the bazaar at Bahram Chah, which served as a Taliban logistics base along the southwestern border between Afghanistan and Pakistan. Task Force 210, a conglomerate of British and Afghan special operations forces, comprised the assault element. As the operation's main effort, Task Force 210's initial mission was to insert under the cover of darkness and seize the bazaar.

The 1st Light Armored Reconnaissance Battalion (1st LAR/Task Force Highlander) served as the security element. Its role was to conduct a simultaneous road march to Bahram Chah, to link up with Task Force 210 at sunrise, and then to secure the objective area by cordoning off villages to the northeast and southwest of the bazaar. This would enable Task Force 210 to focus on searching for contraband and intelligence rather than fighting off insurgents.

The support element was primarily drawn from other I Marine Expeditionary Force (I MEF) units. It included a platoon from 3d Combat Engineer Battalion, which accompanied Task Force Highlander and helped negate obstacles in the convoy's path. A groupment of light howitzers and rocket artillery from 1st Battalion, 11th Marines, eliminated enemy fighting positions along the route and in the objective area. Marines from Combat Logistics Battalion 3's Landing Support Platoon and Motor Transport Bravo Company provided food, water, fuel, and ammunition as well as recovered and repaired damaged vehicles during the operation. Marine Wing Support Squadron 373 established a forward arming and refueling point. Operating from this location, pilots and aircrews from Marine Heavy Helicopter Squadron 466, Marine Medium Tiltrotor Squadron 365, and Marine Light Attack Helicopter Squadron 369 provided logistical, medical evacuation, and close air support to the ground forces. Close air support included the Marine Fighter Attack Squadron 232, flying from the Kandahar International Airport, and U.S. Air Force Rockwell B-1 Lancer bombers, Fairchild Republic A-10 Thunderbolt attack bombers, and Lockheed AC-130 Spectre gunships. U.S. Army Sikorsky UH-60 Black Hawk helicopters were also on hand in case medical evacuations were required.

The raid was an unmitigated success and, according to Major General Richard P. Mills, commander of I MEF and RC-SW, it had a significant impact on the enemy's warfighting ability. In addition to killing between 15 and 25 insurgents, Coalition forces seized nearly 27 metric tons of ammonium nitrate, 60 cases of .50-caliber machine gun ammunition, 22 IEDs, and numerous artillery shells, automatic weapons, and assorted ammunition, as well as 40 kilograms of opium, 500 liters of acid, and 2,000 kilograms of precursor chemicals used to refine narcotics. No Marines were injured during the operation.

Over the next two days, the support and then the assault elements returned to their home stations.[111]

The raid on Bahram Chah was considered an unmitigated success. While no Marines had been injured, official press releases reported that Coalition forces had killed between 15 and 25 insurgents. Reports also acknowledged that those numbers did not include dead or wounded fighters that the enemy might have retrieved during the prolonged action.[112] Other sources, which did account for those losses, estimated that the number of enemy dead might have been more than 70.[113] In addition to seizing nearly 27 metric tons of ammonium nitrate—enough to arm 2,000 IEDs—the raid force also recovered 60 cases of .50-caliber machine-gun ammunition, 22 IEDs, and numerous artillery shells, automatic weapons, and assorted ammunition. Drug-related materials confiscated included 40 kilograms of opium, 500 liters of acid, and 2,000 kilograms of precursor chemicals used to refine narcotics.[114]

In November 2010, 3d LAR, under Lieutenant Colonel Kenneth R. Kassner, relieved Task Force Highlander. As Lieutenant Colonel Leonard had

Defense Imagery 379624, photo by SSgt Jeremy Ross

Marines with 3d LAR scan the hills of the Bahram Chah valley for an insurgent mortar spotter during Operation Rawhide II on 16 March. Throughout the four-day raid on the key insurgent trafficking hub on the Pakistan border, insurgent spotters attempted to creep over the hill to call in fire on Marine positions.

prophesized, Coalition forces returned to Bahram Chah in March 2011. The 3d LAR spent three days in the infamous border town during Operation Rawhide II. The Marines destroyed the bazaar, killed about 50 insurgents, and sustained minimal casualties.[115]

At the end of March 2011, Major General Mills turned command of RC-Southwest over to Major General John A. Toolan. After returning to the United States, Mills described Coalition efforts to deny sanctuary to the enemy and to "fight him on ground of our choosing" as having a "significant impact."[116]

> In a series of fairly aggressive movements up around the Sangin area and south against the border in Bahram Chah, we kept pressure on his lines of supply, kept pressure on his forces as they really tried to move to find areas to rest and relax in. We had some significant gains against him in that way and began to see some of his forces begin to crumble. As the supply lines were cut, we got excellent intelligence that showed us it was impacting him in the fighting holes. He was running out of money. He was running out of equipment, he was running out of ammunition. We saw that in things like him digging up old IEDs to attempt to reuse them on the battlefield. Things he would not have done had he had a warehouse full of them sitting somewhere. We saw him saving his expended ammunition in order to repack his rounds. Things like that showed that he was having supply difficulties. We had very good intel that showed his subordinate commanders were also selling

Defense Imagery 268243, photo by Cpl Jeremy Fasci

MajGen Richard P. Mills and SgtMaj Micheal P. Barrett uncase the unit's colors during a transfer of authority ceremony at Camp Leatherneck, Helmand Province.

personal equipment, such as cars in order to pay their troops. Again, something he would not have done had he had access to the resources that he had at one point in time. So I think the pressure on him, the constant pressure, and the ability to fight him on ground of our choosing had a significant impact on his success.[117]

Epilogue

Despite those gains, Bahram Chah continued to serve as a Taliban logistics center in southwestern Afghanistan. One factor contributing to the site's staying power was the enemy's ability to seek sanctuary in nearby Pakistan. Major General Toolan lamented, ". . . it's like I can't shut the water off, I just keep mopping up the floor. If I could turn the water off in Pakistan, it would be a lot better."[118] The Pakistani Army's XII Corps was of little help on the southern side of the border because its government was preoccupied with a separatist movement in Balochistan, Pakistan.

Another factor was the aggressive withdrawal of American forces from Afghanistan, which began in July 2011 and then accelerated in 2012.[119] As strategic priorities shifted, Task Force Leatherneck focused on training, advising, and turning operating areas over to Afghan security forces.[120] Major General Charles M. Gurganus, who followed Major General Toolan as commander of RC-Southwest, defined success during his tenure "as setting the conditions for the Afghans to take over their own security, their own government, and then they have an opportunity to decide what to do with it."[121] Afghani Lieutenant Colonel Mohammad Rasul Qandahari, who commanded an Afghanistan Border Police battalion situated along the southern Helmand River, conceded that he and his men would need more manpower and larger weapons before they could hope to push southward into enemy territory.[122]

Note

Marine Corps History 1, no. 1 (Summer 2015): 6–22. Reprinted with permission from *Marine Corps History*.

About the Author

Colonel Nathan S. Lowrey, USMCR (Ret), is a historian with the Joint History Office. Colonel Lowrey previously served the History Division for a number of years as a historian, deputy director, and as a member of its Individual Marine Augment (IMA) detachment. He is the author of the acclaimed History Division monograph *From the Sea: U.S. Marines in Afghanistan, 2001–2002* (2011).

Endnotes

1. Associated Press, "Rumsfeld Declares Major Combat Over in Afghanistan," Fox News, 1 May 2003.
2. Thomas H. Johnson and M. Chris Mason, "Understanding the Taliban and Insurgency in Afghanistan," *Orbis* 51, no. 1 (Winter 2007): 72, 81–82, 84.
3. Jeffrey A. Dressler, *Securing Helmand: Understanding and Responding to the Enemy,* Afghanistan Report 2 (Washington, DC: Institute for the Study of War, 2009), 7, 33–34.
4. Ann Scott Tyson, "Marines' Afghanistan Plan Sparks Debate," *Washington Post,* 12 October 2007, A-14.
5. ISAF Joint Command-Afghanistan, "Regional Command Southwest Stands Up in Afghanistan," Resolute Support News, 14 June 2010.
6. Dressler, *Securing Helmand,* 13; I MEF Navy Unit Commendation recommendation, 26 July 2012 (Military Awards Branch [MilAwdsBr], Headquarters Marine Corps [HQMC], Quantico, VA); and Sgt Stuart R. Sanford intvw with Maj Beth Wolney, 8 February 2011 (Oral History Collection [Oral HistColl]). All oral history materials cited in this work can be found at the Gray Research Center (GRC), Quantico, VA.
7. MajGen Richard P. Mills intvw with Dr. Kimberly Kagan, 2 May 2011 (Oral Hist Coll).
8. "Largest Drugs Bazaar Raided," *PakTribune,* 31 May 2005; and Waheed, "Afghan Forces Destroy Country's Largest Drug Bazaar," *Afghan Warrior* (blog), 5 June 2005.
9. Ron Synovitz, "Afghanistan: Police Chief Says Smugglers Helped Taliban with Ambush," Radio Free Europe/Radio Liberty, 11 October 2005.
10. "Afghanistan Raids Drug Bazaar Near Pakistan Border," Voice of America News, 11 April 2006; Daniel Cooney, "U.S. Warns Afghanistan Violence Will Worsen," ABC News, 3 April 2006; and Reuters, "Afghans Seize Four Tonnes of Drugs, Arrest Seven," Afghanistan News Center, 12 April 2006.
11. "Special Boat Service Operations—Hunting Mullah Dadullah," Elite UK Forces, 12 May 2007.
12. "Seventy Taliban Said Killed in Afghanistan," Reuters, 15 October 2008; "Up to 70 Taliban Killed in Afghan Strike: Local Official," Agence France-Presse, 15 October 2008; and "70 Taliban Killed in Afghan Strike," *Daily Times,* 16 October 2008.
13. Sebastian Abbot, "Taliban Sneak Past Marines in Southern Afghanistan," Associated Press, 16 December 2009.
14. Maj John R. Bitonti II intvw with Col N. S. Lowrey, 7 February 2011 (Oral HistColl); Maj John R. Bitonti II award, 2 December 2010 (MilAwdsBr, HQMC, Quantico, VA); and Sgt Dean Davis, "Charlie Company Marines Travel Far and Wide to Reach Afghan Locals," *Marine Corps News,* 4 June 2010.

15. Capt Adrian B. Hascamp intvw with Col N. S. Lowrey, 7 February 2011 (Oral HistColl); LtCol Scott D. Leonard intvw with Col N. S. Lowrey, 8 February 2011 (Oral HistColl); Capt Adrian B. Hascamp award recommendation, 2 December 2010 (MilAwdsBr, HQMC, Quantico, VA); and LtCol Scott D. Leonard award recommendation, 6 June 2011 (MilAwdsBr, HQMC, Quantico, VA).
16. ISAF Joint Command Headquarters, "Joint Patrol Discovers 725 kg of Hashish," 22 March 2010.
17. "June 2 Airpower Summary: KC-135s Provide Fuel Support," *U.S. Air Force News*, 4 June 2010.
18. Thomas Harding, "Special Forces Commando Took on Taliban Single-handedly," *Telegraph*, 24 May 2011.
19. Ibid.
20. ISAF Joint Command Headquarters, "IED Factory, Materials Destroyed in Southern Helmand," 21 September 2010.
21. Marine Medium Tiltrotor Squadron 365 (VMM-365) Command Chronology, hereafter ComdC, October 2010 (GRC, Quantico, VA), Part 2, 3.
22. Leonard intvw.
23. 1stLt Thomas F. Clauss III intvw with Col N. S. Lowrey, 7 February 2011 (Oral HistColl); and Capt Nicholas S. Rapkoch intvw with LtCol R. S. Sellards, 7 February 2011 (Oral HistColl).
24. Capt Nicolas S. Rapkoch award recommendation, 27 December 2010 (MilAwdsBr, HQMC, Quantico, VA).
25. Leonard intvw.
26. Ibid.
27. Rapkoch intvw.
28. 1st MarDiv ComdC, 16 October–15 November 2010 (MCHC, Quantico, VA), Section 2, Fires; and Rapkoch intvw.
29. 1st MarDiv, Task Force Leatherneck After Action Report for Operation Enduring Freedom 10.1–11.1, 24 February 2011 (GRC, Archives, Quantico, VA), 68.
30. LtCol Adolfo Garcia Jr. intvw with Col N. S. Lowrey, 9 February 2011 (Oral HistColl).
31. 1st Battalion, 11th Marines ComdC, 1 May–21 November 2010 (Marine Corps Heritage Center [MCHC], Quantico, VA), Part 2, 6–44; LtCol Adolfo Garcia Jr. comments on draft ms, 10 March 2015 (MCHC, Quantico, VA); and Garcia intvw.
32. Maj David J. Grabow award recommendation, 14 February 2011 (MilAwdsBr, HQMC, Quantico, VA); and 1st Battalion, 11th Marines ComdC, 35–39.
33. Ibid., 38.
34. Garcia intvw.
35. Bitonti intvw; and LCpl Adam B. Ramirez intvw with Col N. S. Lowrey, 8 February 2011 (Oral HistColl).
36. Garcia intvw; and 1st Battalion, 11th Marines ComdC, 23, 25.
37. Leonard intvw.
38. Garcia comments on draft ms.
39. Bitonti intvw; and GySgt Aaron J. Abrams intvw with Col N. S. Lowrey, 8 February 2011 (Oral HistColl).
40. 1st MarDiv ComdC, Part 2, Fires.
41. Abrams intvw; and GySgt Aaron J. Abrams award recommendation, 2 December 2010 (MilAwdsBr, HQMC, Quantico, VA).
42. Maj Jon A. Custis award recommendation, 15 March 2011 (MilAwdsBr, HQMC, Quantico, VA).
43. Cpl Derek B. Carlson, "Afghan, Coalition Forces Kill Insurgents; Seize Drugs, Weapons during Raid," *Marine Corps News*, 1 November 2010.
44. Leonard intvw; and Maj Jon A. Custis intvw with Maj Beth Wolney, 7 February 2011 (Oral HistColl).
45. Cpl Paul D. Zellner, "Wrecker Crew Keeps Convoy Rolling during Operation Steel Dawn II," Defense Imagery, 13 November 2010.
46. 1st Battalion, 11th Marines ComdC, 13, 23; and Garcia intvw.
47. Cpl Paul D. Zellner, "Marines Provide Logistics Support to Coalition Forces during Operation Steel Dawn II," Defense Imagery, 11 November 2010.

48. Leonard intvw.
49. Carlson, "Afghan, Coalition Forces Kill Insurgents."
50. Image courtesy Defense Imagery (337483) in Carlson, "Afghan, Coalition Forces Kill Insurgents"; and VMM-365 ComdC, 3.
51. Carlson, "Afghan, Coalition Forces Kill Insurgents."
52. Abrams intvw; and Abrams award. On 23 and 26 October 2010, Task Force Leatherneck executed dynamic strikes against compounds associated with the production of IEDs, destroying buildings and their contents, according to the 1st MarDiv ComdC.
53. Rapkoch intvw; and Abrams intvw.
54. Ibid.
55. Clauss intvw; and MSgt Curtis C. Gregory intvw with Col N. S. Lowrey, 8 February 2011 (Oral HistColl).
56. Rapkoch intvw; and Ramirez intvw.
57. Ibid.
58. Gregory intvw; and Custis award.
59. Maj David J. Grabow intvw with LtCol R. S. Sellards, 9 February 2011 (Oral HistColl); Garcia intvw; and Grabow award.
60. Leonard intvw; and Mills intvw.
61. Abrams intvw.
62. Rapkoch intvw; Clauss intvw; Abrams intvw; and Rapkoch award.
63. Rapkoch award; and Rapkoch intvw.
64. 1st Battalion, 11th Marines ComdC, 25; Rapkoch award; and Rapkoch intvw.
65. Rapkoch award; Garcia intvw; and Leonard intvw.
66. 1st Battalion, 11th Marines ComdC, 36.
67. Rapkoch intvw.
68. Ramirez intvw.
69. Sanford intvw.
70. Mills intvw.
71. 1stLt Michael D. Wright award recommendation, 2 December 2010 (MilAwdsBr, HQMC, Quantico, VA).
72. Leonard intvw; Bitonti intvw; Gregory intvw; and Rapkoch intvw.
73. Rapkoch intvw; Clauss intvw; Abrams intvw; Garcia intvw; and 1st Battalion, 11th Marines ComdC, 39.
74. Bitonti intvw; and Rapkoch intvw.
75. Mills intvw.
76. Bitonti intvw; and 1stLt Jonathan R. Walaski award recommendation, 2 February 2011 (MilAwdsBr, HQMC, Quantico, VA).
77. Walaski award.
78. Bitonti intvw.
79. Gregory intvw.
80. Abrams intvw.
81. Bitonti intvw; Gregory intvw; and Abrams intvw.
82. Abrams intvw.
83. Bitonti intvw.
84. Sanford intvw.
85. Abrams intvw.
86. Sanford intvw; 1stSgt Jon D. Jerome intvw with LtCol R. S. Sellards, 7 February 2011 (Oral HistColl); and 1stSgt Jon D. Jerome phone conversation with author.
87. Jerome intvw.
88. Leonard award; Rapkoch intvw; and Jerome phone conversation.
89. Custis award.
90. Rapkoch award; and Rapkoch intvw.
91. Jerome intvw.
92. Grabow intvw.
93. Jerome intvw.
94. 1stLt Buck A. Bradley award recommendation, 23 November 2010 (MilAwdsBr, HQMC, Quantico, VA).
95. MajGen Richard P. Mills, "Major General Richard Mills on Afghanistan" (lecture, Carnegie Endowment for International Peace, Washington, DC, 27 April 2011).
96. Clauss intvw; Rapkoch intvw; and Leonard intvw.
97. Clauss intvw; Rapkoch intvw; and Leonard intvw.
98. Gregory intvw; and Clauss intvw.
99. Bitonti intvw; and Abrams intvw.
100. Abrams intvw.
101. Bitonti intvw; and Leonard intvw.
102. Bitonti intvw.
103. Ibid.

104. Abrams intvw; and Gregory intvw.

105. Rapkoch intvw.

106. Bradley award.

107. Bitonti intvw.

108. Clauss intvw; and 1st Battalion, 11th Marines ComdC, 36.

109. Rapkoch intvw; and Abrams intvw.

110. Clauss intvw.

111. Leonard intvw.

112. "Officials in Afghanistan Provide Details of Operations," DOD News, 1 June 2010; and ISAF IJC Headquarters, "Afghan, Coalition Forces Wrap Up Helmand Clearing Operation," DOD News, 8 November 2010.

113. Leonard award.

114. Carlson, "Afghan, Coalition Forces Kill Insurgents."

115. Sgt Jeremy Ross, "3rd LAR Strikes Key Insurgent Border Hub during Operation Raw Hide II," II MEF Marine Expeditionary Force, 19 March 2011.

116. Mills intvw.

117. Ibid.

118. Jim Garamone, "Pakistan, Corruption Affect South, West Afghanistan," American Forces Press Service, 24 April 2012.

119. CNN Wire Staff, "Afghan Withdrawal More 'Aggressive' than Military Proposed," CNN News, 24 June 2011; Chris Carroll, "As Marines Leave Helmand, Focus Turns to Training Those Left Behind," *Stars and Stripes*, 11 July 2012; and Dan Lamothe, "Taliban Make Gains as Marines Take Backseat," *Marine Corps Times*, 29 October 2012.

120. Robert Burns, "Afghanistan: Marines to Wind Down Combat in 2012," *Huffington Post*, 26 November 2011.

121. Dan Lamothe, "Marine General: Lethal Aid Entering Helmand Is Top Concern," *Military Times*, 18 April 2013.

122. Ibid.

LCpl Kyle Carpenter in Helmand Province, Afghanistan.
Defense Imagery 140609-M-AB123-006

"It's Your Medal": Read Kyle Carpenter's Inspiring Speech to Marjah Vets

by Hope Hodge Seck

Marine Corps Times, February 2015

On 13 February 2010, Marines assaulted the Taliban hub of Marjah in Southern Afghanistan in what Coalition leaders would title Operation Moshtarak, from a Dari word meaning *together*, or *joint*. The attack would be the largest military operation since the start of the war in Afghanistan nine years earlier. It would take longer than expected and take a greater toll in casualties than anticipated.

According to a tally by Marjahveterans.com, 78 U.S. and British troops would lose their lives in Marjah and in support of Operation Moshtarak. But the operation would also be a success. Coalition forces ultimately flushed out insurgents from their stronghold and returned greater safety and stability to the region's population.

From the Battle of Marjah also came one of the Marine Corps' greatest stories of heroism out of Afghanistan. Medically retired Marine Corporal Kyle Carpenter would receive the Medal of Honor for throwing himself on a grenade to shield a friend and fellow Marine from the blast during the 2010 Marjah assault.

The Speech

On 13 February, the five-year anniversary of the launch of Operation Moshtarak, Carpenter addressed several hundred Marjah veterans at the National Museum of the Marine Corps. Here, as accurately transcribed as possible from a recording, is his deeply moving and personal speech:

> With this short amount of time I have to speak to you tonight, I couldn't possibly sum up the historical battle of Marjah.
>
> I am comforted, though, by the fact that the men in this room don't need a summary because you were right there beside me. You felt the incredible heat of a 100 percent humidity day and the cool waters of a muddy canal. You felt the weight of 100 pounds of gear, ammo and water at your back, the weight of knowing as Marines we are and forever will be the first line of defense for our loved ones, our nation and above all, freedom.
>
> I stand here today extremely proud of you all. I'm proud of the job you did in the face of what most cannot even fathom. I am more than honored to call you friends, fellow Marines and brothers. You stand as an example for others and for what's best for not only our nation but the rest of the world.
>
> The United States military member is a beacon of hope in dark places for suffering people around the world. Many have paid the ultimate sacrifice. Many gave their limbs to help people have lives free of oppression and full of freedom and prosperity.
>
> Even though there are dark days and have been dark days since our deployment, and long nights, remember what [U.S. Army] Gen. [George S.] Patton said: "It is foolish and wrong to mourn the men who died; rather, we should thank God that such men lived."
>
> Be proud of who you are. Be proud of what you did in that country. You are alive today and have been blessed with this opportunity of life. Don't waste it. Live a life worth living, full of meaning and purpose, and one that will make the fallen who are looking down on us proud.
>
> For the families of the fallen, Blue Star families, active duty families, retired, and all military families and service members, I thank you for your service.* You don't

Official U.S. Marine Corps photo
by Cpl Michael C. Guinto

Retired U.S. Marine Corps Cpl William Kyle Carpenter poses for his official Medal of Honor photo on 18 June 2014.

* The term *blue star family* references the service flag dating back to World War II that was created for families with servicemembers overseas.

hear enough, "I appreciate your sacrifice and what you go through here at home or half a world away deployed."

To everyone here tonight, I thank you for having me. I'm extremely honored to stand in front of you and I'm very humbled that you wanted me to speak to you here tonight.

Marines, I'm proud to have worn the same uniform as you.

Never forget that when no one else would raise their right hand, you did. You sacrificed and became part of our nation's history and our Marine Corps legacy for taking part in the historical battleground of Marjah. Thank you so much. I really do appreciate it.

[Carpenter left the podium, but returned a moment later.]

I really struggled with the idea that I would, not have to, but most likely be encouraged to wear my medal. And I say struggled because, let me just say that I don't want to wear this. I don't like wearing this.

But I do because, you know, if I can inform one person of what we do and what we're about, or what we sacrificed over there, I do it for that. I wear it for all of you.

And I just hope that you know that no matter where I'm wearing it, it's not because I want to. It's putting on a good face, trying to attach something good to the Marine Corps, to contribute and help people understand our side of life, what we go through, what we're about. And everything we've done from past generations until now, the great job we've done to keep our freedom, to keep our men alive.

So I wore it tonight for you. Feel free to come up after and touch it, whatever you like. It's your medal.

Note

Marine Corps Times, 16 February 2015, 21. Reprinted with permission from the *Marine Corps Times.*

About the Author

Hope Hodge Seck was a staff writer with the *Marine Corps Times* from 2013 to 2015.

MSgt Hector Reyes (left), team chief, Georgia Liaison Team, and Sgt Christopher Holm examine the rooftop of a building in a simulated Afghanistan village. Both Marines are with 2d Air Naval Gunfire Liaison Company and are assisting with training the Georgian 32d Light Infantry Battalion.

Defense Imagery 310186, photo by MSgt Grady Fontana

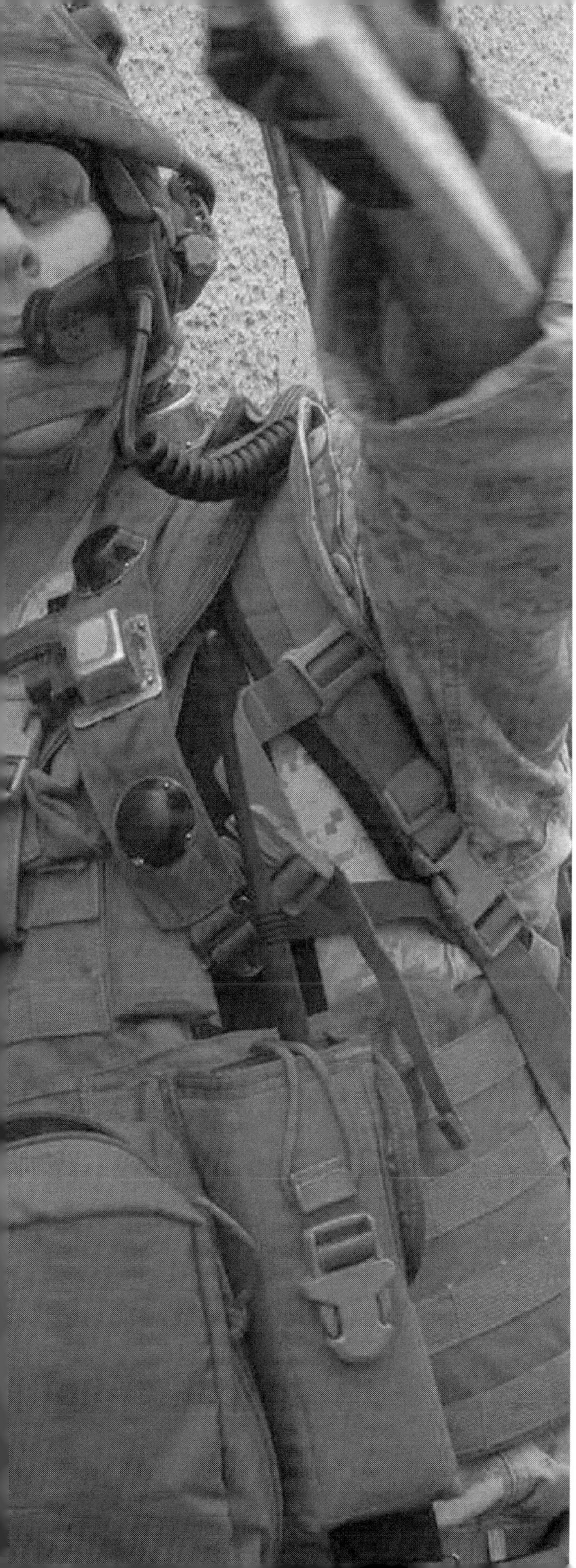

Interview: Captain Jonathan A. Hutchison

by Marine Corps History Division

Marine Corps field historian Major Robert F. Williams III interviewed Captain Jonathan A. Hutchison, 1st Brigade air officer, 2d Air Naval Gunfire Liaison Company (2d ANGLICO), Marine Headquarters Group, on 7 May 2011 in Camp Leatherneck, Helmand Province, Afghanistan, concerning Hutchison's 2010–11 deployment to Afghanistan. The following has been edited from the original transcript for space constraints in this work.

Williams: Would you define what it means to be the 1st Brigade air officer in this environment?

Hutchison: For 1st Brigade, we're kind of partnered with Task Force Helmand so all of our teams are with UK advisor units who are supporting the ANA [Afghan National Army] within Task Force Helmand anywhere from Nad Ali [District] south up to Nez north and with one team in the upper Garesh Valley supporting an ODO [Organizational Development Office] mission.

So, a lot of what I do is briefing Task Force Helmand and its supporting units on what sort of American aviation capabilities exist and how we can utilize those assets to further support their units in the ANA that we're supporting.

Williams: Now, when it comes to—there's a unique aspect here—again, ALCO's [air launch control officer's] job is always to act as the liaison, a translator, if you will. But, when it comes to the UK and the interactions I've had with various British air officers, in particular, rotary, specifically. They have a different kind of mentality, a different mindset and they're not used to being controlled from a ground controller.

Hutchison: Yes.

Williams: I mean, if there are other aspects . . .

Hutchison: There are—

Williams: . . . that you could speak to . . .

Hutchison: There is a very distinct dynamic where their Apaches [attack helicopters] are used to CCAs, close combat attacks, as opposed to a close air support mindset. So, there has been quite a bit of education process on educating the UK, their teams, and their aircrew on how a Marine Corps JTAC [joint terminal attack controller] of FAC [forward air controller] will operate. Their Apaches are definitely used to more of getting suggestions or sort of guidance from guys on the ground as opposed to actually being positively controlled and deconflicted from other assets.*

Additionally, they're not used to sort of a combined arms type of mentality. They're used to being the only fire support asset on station where they're not having to deconflict themselves from mortars and direct fire and that sort of thing. So, there's been a big role in educating them and in teaching them that fire support and deconfliction and what our teams are actually trying to do for them and what added benefit they have with actually a FCT [fire control team] as opposed to just a JTAC.

Official U.S. Marine Corps photo

U.S. Marine Corps Capt Johnathan A. Hutchison.

Williams: Now, are there other—now, I was going to ask to kind of—when you go through the walk-through of your predeployment and deployment experiences if there is a way you could kind of comment on some of the touch points with the Coalition forces. Granted, you opened up with Task Force Helmand. That seems to be the lion's share of your interaction responsibilities here in this AO [area of operations], but as we go back into some of the predeployment training, again, our mentality and our mindset, our culture is dramatically different in some regard to some of the Services that we're trying to support, a.k.a. the whole point behind ANGLICO in the first place. But in this case, the Marine Corps itself has grown well outside our traditional boundaries. We're now running a coalition, which is a whole different animal.

* In this instance, the term *deconflict* means to reduce the risk of collision between aircraft, airborne weaponry, etc., in an area by coordinating their movements.

So, I don't know if it's better to go back and walk through some of the predeployment touch points, or if you'd like and prefer to just deal with it here in the deployment kind of setting, and speak to some of the pronounced interactions.

Hutchison: The issue with predeployment workup is I didn't arrive to the unit until late July. So, the only workup I did was out to the [Fort Irwin] National Training Center [California]. I went out there as the air officer and sort of—that was my first exposure to our ANGLICO teams.

So, essentially that workup was only for me to find out what the team's strengths and weaknesses were prior to deploying and what we could do to turn some of our weaknesses into our strengths prior to this deployment. That was all U.S. Army with a little bit of Coalition air. But, again, the focus was mainly on what our team's strengths and weaknesses were from my perspective.

Following that, the last workup we did was a self-contained currency surface and air shoot in Camp Lejeune [North Carolina]. Again, that was just ANGLICO with some indirect fire assets from 10th Marines and a Marine squadron down at [Marine Corps Air Station] Beaufort [South Carolina]. So, I had very little Coalition interaction prior to coming to Task Force Helmand due to my late arrival at the unit.

Williams: Well then, coming into the deployment itself, once you hit the AO, had you had prior—I mean, when you were mentioning your background, you mentioned the two [McDonnell Douglas F/A-18] Hornet deployments. Was there any opportunity for you to have Joint Forces exposure?

Hutchison: There was, but not UK specific. I had done some exercises on the aviation side of the house with [the] Thais, with the Singapore Air Force, with the Philippines, but not necessarily with the UK and definitely not from a ground FAC perspective; always from an air centric perspective. There was a big learning curve once we got to Lash[kar Gah], was where I was first based out of at the Task Force Headquarters there. There was a steep learning curve on getting up to speed on how the UK operated and the differences between the UK and the Marine Corps. There is a big police training effort out of Lash[kar Gah], but that's also where Task Force Helmand's headquarters is. And when we first arrived, that is where 1st Brigade Headquarters was located at.

Williams: Well, given that you had as great an opening in terms of the predeployment side of the house, you sort of hit the ground running the minute you got to Lash[kar Gah]. Are there some things you could kind of speak to that would kind of define the deployment in this kind of regard? Just experiences or little . . .

Hutchison: When we first arrived to Lash[kar Gah], I was immediately assigned to go augment one of our firepower control teams that was working with a UK-only team in Nez south patrols platoon. So, my first month of the deployment was not necessarily in an air officer role for the entire brigade. It was spent with a UK maneuver element inside of a compound providing them with fire support capability. So, from learning how the UK operates, actually how their ground forces—that was actually very, very beneficial. You spend the first month with the maneuver element and got to see how their planning process actually works on the ground, what sort of concerns a ground commander has, and how his train of thought may be a little bit different from a Marine Corps ground force commander.

Also got a lot of experience directly working

with UK Apaches. So, I got to see first-hand the different information they were looking for and different information that they required, as opposed to: one, being a Marine aviator, what I was expecting to get from a JTAC; and then also, from doing the bulk of my workup training with Marine assets and what they were looking for from me as a JTAC. So, that was extremely beneficial to first-hand knowledge there and then, after that month, to bring that back and pass that information on to my teams and kind of give some guidance and some TTPs [tactics, techniques, and procedures] that I learned while I was out there to help further their success for the next six months of our deployment.

Williams: One of the interesting things—and again, I mention to the room before that I also interview our Coalition partners to understand more about what they see in terms of the Marine experience given they're a part of it. But, the comments are often that some of our mentality, some of our culture is a little hard to take; the "get it done" kind of mentality. And the British, especially when it comes to the Marine Corps planning process, are approached while the methodology makes sense to them, but they have a different way that they go about making decisions.

Hutchison: Yes.

Williams: Was there any exposure or interaction you had on that—while you learned how to work the UK Apaches, were you also learning a little bit of how to work the UK officer and the enlisted corps?

Hutchison: Yes. There was a wide or eye-opening experience with those guys, seeing their different approach to the planning process and some may say the lack of approach to a planning process; much more laid back, much more, not necessarily "these are the wickets we need to hit during the process" as more of a "well, what is our"—

Williams: It's collegial.

Hutchison: Yeah, exactly. Yes, sir. That's 100 percent correct.

Williams: So, how did you react to that? [Laughing]

Hutchison: There's a lot of, especially in the liaison role, is adapting to what they're looking for but also taking the key elements of our planning process and making sure we're hitting those required elements and kind of molding our process to how they're used to operating and to bring out the pluses and minuses of both systems to, in the end, ensure that we have a good product prior to stepping out the door.

Williams: You mentioned that was one eye-opener. I mean, is there a way to describe it or at least put it in some kind of longitudinal perspective?

Hutchison: To be honest with you, the entire month was pretty eye-opening. I mean it seemed, especially from a—from the JTAC perspective, prior to stepping out the door, is that there wasn't much of a fire support plan, integrated plan. It was more of, "Hey, we're going to leave tomorrow on this and go seize this compound and we'll just—we'll ask for—if we need air, we'll just come up and ask for an immediate request—an immediate, priority request for aviation support." It wasn't so much preplanned targets, integration of surface fires. It was more of, "Well, if we need it, we'll just ask for some immediate air support. We'll figure it out from there."

So, getting them to understand the process of, "Well, if you tell me, 'Hey, where are we going? What are some of the historical firing points? What is the enemy's composition out there? What is the commander's intent for fire support? What are his—what is his target priority list?' Those sort of things. You give me that information, I can develop you a much more responsive and prepared fire support plan as opposed to just, 'Well, we'll figure it out as things sort of happen on the battlefield'." I mean everyone understands that no plan survives contact with the enemy, but it's nice to have a plan to deviate from as opposed to already going in and planning on deviating.

Williams: And I appreciate you adding some insight and some color to that because, well, we're not trying to denigrate anyone, it's just we've learned an approach toward doing business and a key facet of the role that you played here is to help translate that back and forth.

Hutchison: Right.

Williams: But sometimes in shaping—the shaping operations of our audience, the UK experience has been very interesting. And they openly comment to it.

Hutchison: Right.

Williams: They recognize— I mean I think sometimes they even have fun with this deployment to a degree because we're an easy group to bait.

Hutchison: I would agree with that. They enjoy that tremendously.

Williams: [Laughing] But, when it came then to—you mentioned the first month was a bit of an eye-opener, beyond that point, as you—did you get out of the FCT team and then start becoming more and more—

Hutchison: At that point, in mid-December, I went back to brigade headquarters. At that time, though the brigade headquarters was in the process of transitioning from Lash[kar Gah] at Task Force Helmand Headquarters down to where we currently are at Camp Tombstone and Shorabak, in support of the 3/215 Corps, Afghan National Army.

Williams: Is it direct support?

Hutchison: It is. Absolutely.

Williams: Now, I'd imagine that's a dramatically different kind of—

Hutchison: That was—it's dramatically different. In conjunction [with] moving our brigade headquarters to Tombstone, we also moved one of our SALTs [supporting arms liaison teams], SALT Bravo, down to Nad Ali South, with the SALT headquarters being at [Forward Operating Base] Shawqat, which is Nad Ali South headquarters. And then two fire power control teams; one at Kalang and one at, initially at— They just recently located to Samsor, which are—and those units are in direct support of the ANA Tolay that's down in Nad Ali South.*

Williams: Now, given that we've commented to what it's like to work with the Brits in your capacity, are there some things and some observations that you can speak to with respect to getting starting and interacting with the Afghan National?

* Within the ANA structure, a tolay is a company-size unit.

Hutchison: The Afghan National Army I would say that at times has been very frustrating. But, I was actually—there has been surprising amount of—they're not as bad as I thought they were going to be, for lack of a better word. There are—their senior—some of their senior leadership, talking about the kandak commanders and tolay commanders, are actually pretty good. And they obviously have a vested interest since this is their home country in ensuring the security of their country. The issue is we come here for six, seven months, maybe a year at a time, where this is their life. So I rip out of here in June. They are still going to be fighting this fight. So, some of their—when it comes across as lazy, is not necessarily, in my opinion, the case. It's more of, "Well, you're leaving in seven months, but I'm going to be here for the rest of my life. So, I am going to take—I'm not going to rush to make this happen." There's more of a, "Well, let's think this through. Let's see the long-term effects. Let's see what's going to happen because when you go home in seven months, you're going to get some R&R and some rest and that sort of thing where I'm going to continue to be here fighting for Afghanistan." So, I think that often gets misconstrued as laziness, where I don't think that that is necessarily the case.

Williams: Well, I think that's one of the aspects of the COIN doctrine and the COIN fight as a whole in that the Marine Corps, while we're often referenced as the enemy, understanding and appreciating the need for—we get—definitely get the commander part, but the not commander piece has applicability, value, and I think the getting it part is often wrapped around our ability to show restraint and making sure that any fires that are brought to bear are done in an appropriate fashion. I'd like to give you an opportunity to comment to that, if you'd like to.

Hutchison: Yeah, to be quite honest with you, a lot of our teams that have—SALT Bravo in particular, who's down there with the ANA has not necessarily been a employment of fires lesson. It has been a teaching and mentoring role of the ANA, of teaching them fire support, teaching them battle tracking, teaching them some combat first aid. All of that in order to make their unit and their tolay a little bit more self-sustaining and improve their overall readiness and how they're able to support their security of Afghanistan.

So, I'm trying to think off the top of my head. I don't believe that any of our teams that I can think of right now have actually connected and employed fires in an offensive manner in support of ANA-led operation or patrols. It's been more of mentoring and advising. Our kinetic fires and employment have been with our teams that are partnered with the British maneuver units.

Williams: Are there other challenges that have kind of come out in this kind of an environment you'd like to speak to, especially given the responsibilities that you had?

Hutchison: Because obviously working with the ANA, there's obviously the language barrier, obviously the cultural barrier, some of the differences between the Afghan culture and the religious aspects that they have.

I can remember one particular incident and—where a request for a call for fire came in during the Muslim prayer time. And essentially, the request was going unanswered because all of the ANA mortar men were conducting their Muslim prayer session. So, that was a very interesting wide—very interesting time and a learning experience, to be quite honest with you.

At the time, it probably was not handled in the best light. But in the long run, I think the message got across that, while there has to be a balance between our religious faith, their Muslim religious faith, and their culture. There's also some of their ANA soldiers that they are supporting that are out on the ground and need their help, so we can't just drop everything during our call for prayer. We have to still be ready to support the guys on the ground.

Williams: Now, I think this culture has kind of exposed us to—one, on the Marine side, it's far more expeditionary in nature, which you're definitely more inclined to participate in. But, it's more familiar.

Hutchison: Speaking of expeditionary, there also was another—we did a couple ANA brigade-size operations and it was very— We pride ourselves in the Marine Corps of being expeditionary and mobile and light and fast and all those sexiness. But, it was very interesting to see the amount of gear that our Marines were carrying compared to the average amount of gear that an Afghan was [carrying]. He was much more mobile, much more light, much more capable of actually going on a three- to four-hour patrol, helicopter inserted, helicopter extract than the amount of gear that our average Marine or British advisor was carrying. So, that was actually pretty eye-opening.

Williams: And the only—well, I was going to say as an infantry officer and aviator, coming from an aviator, that's a very interesting comment. But, all that facetiousness aside, we have a lot more requirements. You know, they have to wear PPE [personal protective equipment]. You have to— I mean I look at a Marine rifle squad today and I don't recognize half the— They've got more gear than you can shake a stick at. EM [electronic measures] gear. The CID [combat identification] threat out of here is considerable and it's definitely telling. But, in interviewing a lot of the Marines associated with those foot movements, they put more confidence in their eyeballs than they put into the—

Hutchison: Yes. And Talons—

Williams: AOs.

Hutchison: And some of it was PPE but a lot of it also was quote/unquote comfort items; maybe another set of camis or an extra warming layer or additional chow or things like that. Where they, the ANA, had actually stripped down much more, a lot of their comfort items to allow themselves to be more mobile and expeditionary, if you will. So, that was—that was pretty darned—

Williams: Well, there's always something that we can learn. I mean, we do pride ourselves on being able to accommodate more of that than some. But I guess it's always irrelevant for any time.

Hutchison: Right.

Williams: Were there any other challenges that presented themselves?

Hutchison: Kind of on that last one of the points originally where Friday is an ANA religious holiday so we're not— There's not much patrolling. Excuse me. Essentially zero ANA patrolling on Fridays, which was a learning experience for some of our guys who hadn't been out here before, but also for the British advisors. They would come up with a patrolling plan, a patrolling matrix that would include a patrol on Friday and the ANA was like, "No, we're not doing that. We don't patrol on Fridays." So, there was a learning experience there to go like, 'Hey, the Fridays are their religious holiday.' We need to respect that. But what we can do is we can use this—what they are willing to do is take classes on Fridays? And we can use it as a rest and refit and planning day for ISAF and Coalition forces to develop the patrolling schedule of the following week. Let's just move that and do that all on Friday when the ANA is not going to plan on patrolling or any movements outside the wire." So that was initially a big challenge, especially wanting to do operations or things of that nature that would maybe span over a Friday.

Williams: I haven't heard anybody else ever comment to that. I mean I have—it hasn't—conversations that have gone down that road but I can see how that could be a—

Hutchison: Yeah, if you go back and just look at our historical sitreps [situation reports], you will see Fridays are—Friday through Sunday, I would say—are generally slow, but Fridays all my guys. No patrols, rest and refit, plan the weekly patrol schedule, conducted this class or something to that effect.

Williams: Now, when it comes to your successor, what kind of issues—and, again, you mentioned that there might be some complexities with respect to identifying specifically who your successor would be or exactly the extent of the role. But if there was—I mean and had an opportunity to speak with, I think it's 5th ANGLICOALIC. Is it 1st ANGLICO?

Hutchison: [The] 1st ANGLICO. I spoke with Major Palihchale [*sic*] who is the 1st Brigade platoon commander. I've also spoken with the 1st Brigade air officer since I'm kind of dual rolling right now, which is actually a buddy of mine that I know. And I would say that [is] their biggest challenge. One, they're bringing a few less bodies than we are. So, initially, where we had some guys in the reserve at the brigade level, their reserve is kind of already being eaten up, falling in on some of the missions that we're on, enduring missions that we already have.

But additionally, the 4th Kandak is slated to take over some battlespace in Lashkar Gah here in mid-June. There's also a couple other Kandaks that are slated under the [ANA's] 3/215 [Artillery] Corps to start taking over some battle space shortly thereafter. So, I would say that their biggest challenge, in my opinion, is going to be what fire support is required for that battle space that the ANA is to now own and the balance between teaching fire support and actually providing fire support to the ANAs is going to be tricky. In addition, where do you prioritize because ANGLICO itself, 1st Brigade, at most I can make 12 teams for 2d ANGLICO, 1st Brigade. Twelve fixed would be the most I could do. Based on their laydown, I think their number is closer to 10 fixed. So, we— Now we have to figure out where we're going to prioritize, which kandak and which tolay and how are we going to be able to provide them with this capability with our limited ANGLICO resources. And then something that it seems that a lot of Task Force Helmand is struggling with is what is the mentor/advisor role going to actually look like. What is that— How is that capability going to flush out and what will the requirement be for fire support in an advising role?

Williams: Well, speaking of advising, what advice did you throw in their way? You kind of commented [that] the challenge is going to be how do you prioritize everything. Given what you know now, what advice would you pass on to them?

Hutchison: I've been— Obviously passing them information on the time schedule for when the tolays and kandaks are going to take over battle space. And we're tried to get a priority from Task Force Helmand and from the 205th Corps on what are their priority kandaks. How do they fall out? And maybe go from the role of SALT Breeze and SALT Bravo again as an example where they're kind of direct support to that kandak that's down in Nad Ali South. Maybe now our SALTs are still direct support to the 205th Corps and when the commander, General Shirin Shah decides, "Hey, 3d Kandak down in Nad Ali South, they're not my priority for the next month or so but the 2d Kandak is going to be my priority," so maybe then I— Maybe then it's easier to move my ANGLICO support from Nad Ali South over to the 2d Kandak to be able to still provide that support for the commander.

Williams: So try and help them problem-solve.

Hutchison: Exactly, and figure out what their priorities are and then maximize our capability based on where their priorities are, as opposed to just keeping a team with one kandak, maybe they're a little bit more fluid within 205th Corps. That's kind of my recommendation the 1st ANGLICO to look at that.

Williams: Now, all in all, what kind of impact—and this is where I like to at least give you the opportunity to comment on the impact to the Marine Corps, be it our doctrines, specifics associated with the ANGLICO community and the ANGLICO orientation; or if you would like to take it so far as the aviation LOS [line of sight], aspects of what you see in this environment that might be relevant. I'm just looking to try and see—you know, coming out of OEF [Operation Enduring Freedom]—what kind of touch points do you think this is going to have on the Marine Corps?

Hutchison: That's an interesting question. I would say our movement toward the ANA and 3/215 has gotten us into a position where we can better build that relationship with the ANA and partner with the ANA for a lasting legacy. But a lot of the teams I keep preaching to is there is a frustration with some of our teams because they came over and, for lack of a better word, wanted to smash things and blow things up. But what I tell them constantly is you could employ a hundred 500-pound bombs and have very little long lasting implications within this COIN fight with the development ANA fire support, teaching them how to employ their own weapons systems and to conduct fire support once the Marine Corps has left Afghanistan, and teaching them some of the basic combat medical capability. That's one of the big ones that they want to learn, that we all take for granted when we go through boot camp and predeployment workups and that sort of thing. So, teaching them that fire support, teaching them that basic medicine, teaching them how to train their ANA soldiers is going to have a much longer lasting impact than any number of kinetic events that you conduct in Afghanistan.

Williams: And you think, in turn, that's also reinforcing some of these things in our mind as well?

Hutchison: Yes, sir.

Williams: Because I think a lot of the interviews I've conducted with GC [ground combat] elements, somewhat just as support elements as well, but—and there's a constant stress on basic fundamentals and the irony of what I was trained to do in Iraq or getting killed here.

Hutchison: That's right.

Williams: So there's a need to focus on the fundamentals and then apply them appropriately in the environment you find yourself in as opposed to constantly superimposing environments all the time.

Hutchison: So, from the air officer or FACT perspective, I tell all our teams it's brilliance in the basics. We go home and you learn—when you first start training, it's a single, maybe even one aircraft, maybe a single section of aircraft on station. That's the only accident you have and you work on prosecuting targets with that one aspect.

But everyone wants to get to this sexy mindset of, "Well, I've got mortars. I've got artillery. I've got a section of fixed-wing aircraft. I've got a section of rotary-wing aircraft. And I'm going to bring to bear all the horns of the dilemma on the enemy." And in honesty, there I can think of one kinetic event that our teams have had and that was probably—that was four months ago where we had multiple assets on station. The vast majority of our kinetic events have been a single asset, whether it's a single section of Apaches or mortars or indirect or artillery. So, it definitely is brilliance in basics and remembering the basic skills that we learned, as opposed to trying to make everything sexy and very dynamic, if you will.

Williams: Would you mind— I mean, I again, when I ask for you to share what you feel like sharing, but that multiple fire support asset example, can you expand upon that a little bit, because I, again, like yourself, I haven't seen too many instances where—

Hutchison: It was December 30th, one of our teams in Nez South partnered with [the] UK only, so no ANA element, went out on a helo in-fill. Shortly after stepping off the helicopter, there was a nonbattle injury. A guy stepped in a canal the wrong way and twisted his ankle—knee. Sorry. Twisted his knee. So, in the process of getting him taken care of and conducting a medevac, there were three to four British soldiers inside of this canal, which happened to be down the direct access of an enemy machine gun mod. So, those nonbattle injuries quickly became battle injuries with the employment of the enemy machine gun down the line.

So, our FST [fire support team] team had immediately got a section of Apaches on station. But prior to that section of Apaches coming on station, they were suppressing the target with mortars and other indirect fire assets as well as the small-arms guys maneuvering onto the battlefield to get into position to engage with small arms under this machine gun position.

Also, you had a "Pedro" who had to come on station to conduct the medevac and he had—two Pedros actually—so now, within the short, probably 5–10 minutes, you went from a nonbattle injury to our JTAC controlling mortars on the machine gun position, a second enemy firing position inside of a compound and now opened up on the maneuver element.* So, he is engaging that compound with Apaches and he is trying to deconflict both of these to allow the Pedro to get into a hasty HLS [helicopter landing site] to extract the multiple casualties that we had. That is probably the most dynamic 1st Brigade has been in in any contact with multiple assets on station.

Williams: It is a very interesting management requirement.

* The term *Pedro* is a nickname/call sign referring to search and rescue helicopters.

Hutchison: Right. Compounding that is, as we touched upon earlier, is now you have a CCA platform who is used to operating as the only show in town.

Williams: Oh, so these were British Apaches?

Hutchison: British Apaches operating as CCA not wanting to—basically questioning restrictions that he's getting from our JTAC. So, now he's juggling all of these fire support patrol measures, trying to prosecute targets, talk the Apaches onto the compound where we're getting fire from, and now he's also having to explain the reason why we're putting these restrictions in place and trying on the fly [to] teach this UK Apache—

Williams: In the heat of battle.

Hutchison: That, "Hey, man, you're not— At the moment, you are not the only show in town. As a matter of fact, you're not even my most important show in town. And if I need to kick you out of the battle space, I will."

That happened in late December and that was actually a very good learning experience for myself and also for JHFA, which is Joint Helicopter Force Afghanistan.

And that actually led to me having a meeting with their CO [commanding officer] and some of their training officers and ultimately going to the UK during this deployment to discuss about how ANGLICO operates inside a task force element's battle space, how a FAC is dif—a Marine FAC is different than a UK FAC, how a Marine close air support asset is different from a UK—

Williams: So you literally went to London or—

Hutchison: I went to Warminster [England] for 10 days in support of [Exercise] Pashtun Horizon III, which is the brigade staff exercise for the commandos that are out here now. And that was in February.

Williams: That's pretty rare. I haven't heard—

Hutchison: Yeah, that was . . . I was the only ANGLICO Marine and there was four or five Marines back from the states and then there were two, sorry, three guys from the MEB [Marine expeditionary brigade]—one MEB planning section that went out there with me. But, they were more in the conducting of the exercise from a higher headquarters standpoint. I was the only Marine that was actually on the Task Force Helmand floor in the UK actually talking about deployment issues and concerns in real-life ways ANGLICO operates in the battlefield and ways that they could improve support for their supporting units.

Williams: Well, that— You've been very courteous and then adding this other aspect brought out two more points that I had not expected so I appreciate the ground that you've covered and answering and addressing the questions I had prepared coming in. But, as I mentioned before, I'd like to give you at least the opportunity to add any final comments that you have or any points that you'd like to make sure [to] get on an historical record before we close the interview.

Hutchison: I think that's about it, sir. I had wanted to touch on the Pashtun Horizon trip like we just did. So, no, I'm pretty happy with that.

Williams: And it's literally called the Pashtun Horizon?

Hutchison: Pashtun Horizon III. So, they do

—that's the third and final Pashtun Horizon exercise in the British brigade's workup. So, that is supposed to be where they're at the run phrase, for lack of a better word. And they— What happens is—

Williams: So, that's like their Enhanced Mohave migrant?

Hutchison: Similar, but it's all staff. There's no actual maneuver elements on the ground. So, there is a—there are guys who just—British soldiers who just got back from Afghanistan within like the last year or so sit in sort of like another building that's adjacent and they inject scenarios that the oncoming Task Force Helmand headquarters has to solve. And so, there is a downed aircraft scenario. There is an attack on a PB [patrol base] scenario. There's all these sorts of different scenarios that, from a staff standpoint, that they have to crack the nut on. But in addition to that, there are obviously JFHA [*sic*] CO is there so that provided me with an opportunity to meet with their oncoming CO and give him a brief on ANGLICO's capabilities and how we operate different which has further allowed us to develop a UK specific [procedures] for 1st ANGLICO.

So, during 1st ANGLICO's RSOI [reception, staging, onward movement, and integration], they will get a representative from JFHA will come and talk to them about UK Apaches and how they operate. Their air officer. The first thing their air officer will sit down with Task Force—with JHFA's training officer and they will hash out TTPs and discuss differences and he'll be able to push that down immediately to his team as opposed to waiting for two months into his deployment.

Also, the—this is interest, I'm glad we're having this—the other thing is one of my biggest concerns with our predeployment workup, at least from my perspective, is we get ad nauseam ROE [rules of engagement] and national caveats tactical directives. But, I always get it and my teams always get it from a U.S. officer, U.S. JAG [judge advocate general]. Operating at Task Force Helmand, I never get it from a British legal officer or from the UK perspective.

So, we've worked with the UK leg and they are developing a package that 1st ANGLICO will get during their RSOI to talk about some of the subtle differences and the way the British interpret some of these directives and documents. And all— A lot of that came from me going to PH [Pashtun Horizon] III and discussing these options.

Furthermore, UK's legal department didn't even realize that they had American fire support teams operating in British battle space. So, just getting that information out to them and letting them know, "Hey, some of your British units and some of your ANA that operate in your battlespace are actually having fires employed by American Marine JTACs."

Williams: [UK] Brigadier [General] George Norton, when I interviewed him, brought something up along this line. It didn't— I wasn't tracking on what he was describing but I wouldn't be too surprised if this was very pronounced. I mean it's one thing to be in a Coalition fight. It's one thing for it to be led by a Marine major general. It's another thing for all the other Coalition partners to know how to function through that fire.

Hutchison: Correct.

Williams: So, I appreciate you going through that example. Is there anything else I've got to shake you on or—

Hutchison: Well, there's probably one more—one

more thing it's brought to light is no one in Task Force—shouldn't say no one. From what I can understand is that no one had really thought through an unpartnered ANA or ANSF patrol who is now out, that's being supported by a ANGLICO JTAC or FAC from a COC [combat operations center]. So, we have an ANA patrol that's out with just ANA or just ANP or mixed, whatever it might be, all Afghan forces. But, they are being supported from a COC with a Marine JTAC. What are— What sort of fire support capability am I allowed to support [them] with? Do I—only in self-defense? Do I have [it] if the ground force commander in the COC has it? Can I extend that to that ANSF patrol that's out? Can I request immediate for close air support? Can I use British or American indirect fire assets to support this patrol?

So, a lot of those questions are currently being answered and some documentation's being developed on that all from ANGLICO being partnered with these units and asking these questions on what we can and cannot do in support [of] ANA inside a task force element.

Williams: I would have thought they'd just been viewed as another Coalition force, even though it's the national force.

Hutchison: There is a SOP that's out that says an ANSF patrol by themselves does not necessarily rate close air support or fires outside of self-defense. So, we're just trying to get some clarification on that.

Williams: Any other points? I mean, I'm not really that surprised because of the unique nature of what ANGLICO is all about. You're, I'm sure during the course of the conversations I'll have with some of the other Marines, things will come up like this.

Hutchison: Oh, definitely.

Williams: I mean these are telling moments in our history just as prominent as maybe [Major] General Richard P. Mills leading Coalition forces encounters for the first time. There's going to be a lot of firsts coming out of here and I'm glad we're able to get at least some of this on the record and appreciate your comments.

Part II
Taliban Hunting, 2011

In the Shadow of the Poppy Harvest by Michael D. Fay.
Marine Corps Art Collection

Throughout 2011, Marine units rotated into Afghanistan and continued to conduct raids and patrols throughout the Marineistan provinces, suppressing the poppy harvest and eliminating Taliban caches and sanctuaries. In March, II Marine Expeditionary Force (Forward) replaced I Marine Expeditionary Force (Forward) as Regional Command-Southwest.

In May 2011, U.S. Navy SEALs raided Osama bin Laden's hiding place in Pakistan, killing the al-Qaeda leader. In June, President Barack Obama announced the end of the Afghanistan surge and the withdrawal of more than 30,000 American troops by the following summer. On 18 July 2011, Marine General John R. Allen took command of the International Security Assistance Force (ISAF) and U.S. Forces Afghanistan (USFOR-A), becoming the senior American commander in Afghanistan.

President Barack H. Obama addresses soldiers, sailors, airmen, and Marines at Bagram Airfield, Parwan Province, Afghanistan, on 2 May 2011. The president visited Afghanistan to sign a historic agreement with Afghan President Hamid Karzai, setting the path for the eventual withdrawal of U.S. forces and increased self-governance by the recognized government of the Islamic Republic of Afghanistan.

Defense Imagery 120502-A-AW125-009

Remarks by the President on the Way Forward in Afghanistan

by President Barack H. Obama

The White House
Office of the Press Secretary
For Immediate Release
22 June 2011
8:01 P.M. EDT

THE PRESIDENT: Good evening. Nearly 10 years ago, America suffered the worst attack on our shores since Pearl Harbor. This mass murder was planned by Osama bin Laden and his al-Qaeda network in Afghanistan, and signaled a new threat to our security—one in which the targets were no longer soldiers on a battlefield, but innocent men, women, and children going about their daily lives.

In the days that followed, our nation was united as we struck at al-Qaeda and routed the Taliban in Afghanistan. Then, our focus shifted. A second war was launched in Iraq, and we spent enormous blood and treasure to support a new government there. By the time I took office, the war in Afghanistan had entered its seventh year. But al-Qaeda's leaders had escaped into Pakistan and were plotting new attacks, while the Taliban had regrouped and gone on the offensive. Without a new strategy and decisive action, our military commanders warned that we could face a resurgent al-Qaeda and a Taliban taking over large parts of Afghanistan.

For this reason, in one of the most difficult decisions that I've made as president, I ordered an additional 30,000 American troops into Afghanistan. When I announced this surge at West Point, we set clear objectives: to refocus on al-Qaeda, to reverse the Taliban's momentum, and [to] train Afghan security forces to defend their own country. I also made it clear that our commitment would not be open-ended, and that we would begin to draw down our forces this July.

Tonight, I can tell you that we are fulfilling that commitment. Thanks to our extraordinary men and women in uniform, our civilian personnel, and our many Coalition partners, we are meeting our goals. As a result, starting next month, we will be able to remove 10,000 of our troops from Afghanistan by the end of this year, and we will bring home a total of 33,000 troops by next summer, fully recovering the surge I announced at West Point. After this initial reduction, our troops will continue coming home at a steady pace as Afghan security forces move into the lead. Our mission will change from combat to support. By 2014, this process of transition will be complete, and the Afghan people will be responsible for their own security.

We're starting this drawdown from a position of strength. Al-Qaeda is under more pressure than at any time since 9/11. Together with the Pakistanis, we have taken out more than half of al-Qaeda's leadership. And thanks to our intelligence professionals and Special Forces, we killed Osama bin Laden, the only leader that al-Qaeda had ever known. This was a victory for all who have served since 9/11. One soldier summed it up well. "The message," he said, "is we don't forget. You will be held accountable, no matter how long it takes."

The information that we recovered from bin Laden's compound shows al-Qaeda under enormous strain. Bin Laden expressed concern that al-Qaeda had been unable to effectively replace senior terrorists that had been killed, and that al-Qaeda has failed in its effort to portray America as a nation at war with Islam—thereby draining more widespread support. Al-Qaeda remains dangerous, and we must be vigilant against attacks. But we have put al-Qaeda on a path to defeat, and we will not relent until the job is done.

In Afghanistan, we've inflicted serious losses on the Taliban and taken a number of its strongholds. Along with our surge, our allies also increased their commitments, which helped stabilize more of the country. Afghan security forces have grown by over 100,000 troops, and in some provinces and municipalities, we've already begun to transition responsibility for security to the Afghan people. In the face of violence and intimidation, Afghans are fighting and dying for their country, establishing local police forces, opening markets and schools, creating new opportunities for women and girls, and trying to turn the page on decades of war.

Of course, huge challenges remain. This is the beginning—but not the end—of our effort to wind down this war. We'll have to do the hard work of keeping the gains that we've made, while we draw down our forces and transition responsibility for security to the Afghan government. And next May in Chicago, we will host a summit with our NATO allies and partners to shape the next phase of this transition.

We do know that peace cannot come to a land that has known so much war without a political settlement. So as we strengthen the Afghan government and security forces, America will join initiatives that reconcile the Afghan people, including the Taliban. Our position on these talks is clear:

Defense Imagery 345818, photo by SSgt Michael Sparks

President Barack Obama makes a surprise visit to Bagram Airfield on 3 December 2010.

they must be led by the Afghan government, and those who want to be a part of a peaceful Afghanistan must break from al-Qaeda, abandon violence, and abide by the Afghan constitution. But, in part because of our military effort, we have reason to believe that progress can be made.

The goal that we seek is achievable, and can be expressed simply: no safe haven from which al-Qaeda or its affiliates can launch attacks against our homeland or our allies. We won't try to make Afghanistan a perfect place. We will not police its streets or patrol its mountains indefinitely. That is the responsibility of the Afghan government, which must step up its ability to protect its people, and move from an economy shaped by war to one that can sustain a lasting peace. What we can do, and will do, is build a partnership with the Afghan people that endures—one that ensures that we will be able to continue targeting terrorists and supporting a sovereign Afghan government.

Of course, our efforts must also address terrorist safe havens in Pakistan. No country is more endangered by the presence of violent extremists, which is why we will continue to press Pakistan to expand its participation in securing a more peaceful future for this war-torn region. We'll work with the Pakistani government to root out the cancer of violent extremism, and we will insist that it keeps its commitments. For there should be no doubt that, so long as I am president, the United States will never tolerate a safe haven for those who aim to kill us. They cannot elude us, nor escape the justice they deserve.

My fellow Americans, this has been a difficult decade for our country. We've learned anew the profound cost of war—a cost that's been paid by the nearly 4,500 Americans who have given their lives in Iraq, and the over 1,500 who have done so in Afghanistan—men and women who will not live to enjoy the freedom that they defended. Thousands more have been wounded. Some have lost limbs on the battlefield, and others still battle the demons that have followed them home.

Yet tonight, we take comfort in knowing that the tide of war is receding. Fewer of our sons and daughters are serving in harm's way. We've ended our combat mission in Iraq, with 100,000 American troops already out of that country. And even as there will be dark days ahead in Afghanistan, the light of a secure peace can be seen in the distance. These long wars will come to a responsible end.

As they do, we must learn their lessons. Already this decade of war has caused many to question the nature of America's engagement around the world. Some would have America retreat from our responsibility as an anchor of global security, and embrace an isolation that ignores the very real threats that we face. Others would have America overextended, confronting every evil that can be found abroad.

We must chart a more centered course. Like generations before, we must embrace America's singular role in the course of human events. But we must be as pragmatic as we are passionate; as strategic as we are resolute. When threatened, we must respond with force—but when that force can be targeted, we need not deploy large armies overseas. When innocents are being slaughtered and global security endangered, we don't have to choose between standing idly by or acting on our own. Instead, we must rally international action, which we're doing in Libya, where we do not have a single soldier on the ground, but are supporting allies in protecting the Libyan people and giving them the chance to determine their own destiny.

In all that we do, we must remember that what sets America apart is not solely our power—it is the principles upon which our union was founded. We're a nation that brings our enemies to justice while adhering to the rule of law, and respecting the rights of all our citizens. We protect our own freedom and prosperity by extending it to others. We stand not for empire, but for self-determination. That is why we have a stake in the democratic aspirations that are now washing across the Arab world. We will support those revolutions with fidelity to our ideals, with the power of our example, and with an unwavering belief that all human beings deserve to live with freedom and dignity.

Above all, we are a nation whose strength abroad has been anchored in opportunity for our citizens here at home. Over the last decade, we have spent a trillion dollars on war, at a time of rising debt and hard economic times. Now, we must invest in America's greatest resource—our people. We must unleash innovation that creates new jobs and industries, while living within our means. We must rebuild our infrastructure and find new and clean sources of energy. And most of all, after a decade of passionate debate, we must recapture the common purpose that we shared at the beginning of this time of war. For our nation draws strength from our differences, and when our union is strong no hill is too steep, no horizon is beyond our reach.

America, it is time to focus on nation building here at home.

In this effort, we draw inspiration from our fellow Americans who have sacrificed so much on our behalf. To our troops, our veterans and their families, I speak for all Americans when I say that we will keep our sacred trust with you, and provide you with the care and benefits and opportunity that you deserve.

I met some of these patriotic Americans at Fort

Campbell. A while back, I spoke to the 101st Airborne that has fought to turn the tide in Afghanistan, and to the team that took out Osama bin Laden. Standing in front of a model of bin Laden's compound, the Navy SEAL who led that effort paid tribute to those who had been lost—brothers and sisters in arms whose names are now written on bases where our troops stand guard overseas, and on headstones in quiet corners of our country where their memory will never be forgotten. This officer—like so many others I've met on bases, in Baghdad and Bagram, and at Walter Reed and Bethesda Naval Hospital—spoke with humility about how his unit worked together as one, depending on each other, and trusting one another, as a family might do in a time of peril.

That's a lesson worth remembering—that we are all a part of one American family. Though we have known disagreement and division, we are bound together by the creed that is written into our founding documents, and a conviction that the United States of America is a country that can achieve whatever it sets out to accomplish. Now, let us finish the work at hand. Let us responsibly end these wars, and reclaim the American Dream that is at the center of our story. With confidence in our cause, with faith in our fellow citizens, and with hope in our hearts, let us go about the work of extending the promise of America—for this generation, and the next.

May God bless our troops. And may God bless the United States of America.

END
8:16 P.M. EDT

Austere living is just part of the expeditionary nature for Marines, and the leathernecks of 3d Squad, 1st Platoon, Fox Company, 2d Battalion, 3d Marines, did not live in the lap of luxury in Afghanistan's Helmand Province.

Photo by Ed Darack

The Everyday Warfighters: Indefatigable Mission Fidelity in the Modern Toil of War

by Ed Darack

Leatherneck, July 2011

"Corporal Adams and his squad are the Marines you want to be with," a sergeant declared to me not long after I dragged my gear off a [Sikorsky] CH-53D Sea Stallion at Forward Operating Base (FOB) Marjah. "They've found more Taliban weapons and explosives caches than anyone around here. They find something, from a couple AK[47] magazines to huge stocks of explosives, every day they go outside the wire." The sergeant paused for a second. "They sniff that stuff out like hungry dogs. They never come back empty-handed."

I had traveled to RC (SW)—Regional Command (Southwest) in southwestern Afghanistan's Helmand province—to witness firsthand the small-unit Marine Corps infantry experience in remote, austere conditions. After dumping my bags, I walked to the combat operations center, wondering where I was ultimately bound, to which unit I would attach and for how long I would be attached. FOB Marjah, a medium-size base on the outskirts of the city of Marjah itself and the headquarters for 3d Battalion, 9th Marine

Photo by Ed Darack

Squad leader Cpl Robert Adams of 3d Squad, 1st Platoon, Fox Company, 2d Battalion, 3d Marines.

Photo by Ed Darack

SAW gunner LCpl Steven Lopez, 3d Squad, 1st Platoon, Fox Company, 2d Battalion, 3d Marines, patrols the outskirts of Marjah.

Regiment, might have been as far into the hinterlands of RC (SW) as I would be able to push for all I knew at that point.

The battalion had deployed three line companies—India, Kilo, Lima (with attachments)—throughout respective locations in the greater Marjah region. So I was surprised when one of 3d Battalion, 9th Marine's staff offered me the opportunity, in addition to India, Kilo, and Lima Companies to embed with Foxtrot Company.

"Fox/2/3," he responded to my puzzled look. "Second Battalion, 3d Marines, temporarily gave us Fox [Company] to help in our part of the AO [area of operations]. [The] 3/9 covers a large area, and while the core region of Marjah is completely secure, the outer areas have activity." I asked about Corporal Robert Adams' squad, the "king of caches," as I had heard another Marine call them. Just which squad were they? And could I embed with them?

"Third Squad, 1st Platoon, Fox/2/3, and, yes, you can." The next day, I climbed inside an M-ATV (Mine Resistant Ambush Protected, All Terrain Vehicle, the replacement for the humvee in many tactical situations in Afghanistan) and passed outside the wire of the base as part of a resupply convoy headed to the area's PBs (patrol bases), the smallest, most spartan of the Marine outposts in Afghanistan. Those convoys, operated by Marines of "Mobile," serve as the lifeline to the troops living in the PBs, dropping off food, supplies, mail, troops—anything and everything for the infantry Marine.

With the ever-present potential for devastating IED explosions on the myriad dirt roadways interconnecting the small outposts, the Marines of Mobile have one of the riskiest jobs in the Marine Corps. Mobile Marines, however, have an array of technology at their disposal to identify and defeat such threats. Despite this, explained

Photo by Ed Darack

SAW gunner LCpl Joey Marshall, 3d Squad, 1st Platoon, Fox Company, 2d Battalion, 3d Marines, hands a SAW to LCpl Steven Lopez during a patrol outside the city of Marjah.

Corporal Griffin W. Sutherland, "The best defense against IEDs is our ability to focus in on telltale signs of emplaced explosives with the naked eye."

Sutherland described a horrific tragedy that had occurred just weeks prior, when a local Taliban sympathizer emplaced an IED made up of more than 100 pounds of high explosives under a dirt road a few hundred yards from the entrance to one of the PBs. One of the man's neighbors, upon discovering his actions, raced to warn an approaching Mobile convoy. But it was too late. A local farmer driving a tractor set off the pressure-plate trigger, causing an explosion that could be felt for miles and instantly killing his two children who were riding with him.

Marines and Navy corpsmen, risking their lives by possibly setting off other devices placed near the detonated IED (a common tactic), raced to save the farmer, who had literally been torn in two. By the time a medevac helicopter arrived just minutes later, however, he had bled out. The irony of the story, Sutherland explained, was that the man and his children were related to the Taliban sympathizer who had emplaced the IED. "We found the tractor's engine over 300 meters from the site of the explosion."

Photo by Ed Darack

Patrolling and establishing a consistency of presence helped the squad in its counter-IED program and its efforts to locate weapons caches. Machine gunner LCpl Steven McArthur, Fox Company, 2d Battalion, 3d Marines, contributes to that presence in a February 2011 patrol outside of Marjah, Helmand Province.

IEDs continue to be the greatest threat to both local civilians and U.S. and Coalition troops alike in Afghanistan, notably in places such as Sangin and areas outside of the city of Marjah, both in Helmand Province. The very best way to defeat IEDs, however, lies not with technology or honed tactics aimed at detecting where such devices have been emplaced, but by denying the enemy the materials to make them in the first place, mitigating the threat completely.

"I always expect that there is something everywhere we patrol," Corporal Adams stated soon after we met. "If you don't expect that there is something out there, then you won't find anything. The enemy is good at hiding explosives, rifles, triggering mechanisms, you name it."

Adams explained that, for him and his squad, the crux of the current fight is smashing the enemy's destructive capability. Specifics of the enemy's tactics are not really necessary to explain, but, broadly, weapons and explosives come from places other than the Marjah region where "runners" cache them for others to use later. "I take it as a personal defeat not if an IED goes off, but if one even gets made and emplaced," he said. "If one goes off, injuring or killing someone, it's an absolute tragedy."

While Corporal Adams' squad has moved around

the area's patrol bases, they called "Typhoon-5" home when I joined them. Typhoon-5, or T-5, ranks as one of the smallest bases in Afghanistan. Surrounded by mud walls, T-5 has barely the footprint of two suburban homesites, maybe 200 feet by 200 feet total. Most of the Marines there sleep in large tents heated by diesel-fueled stoves; a few sleep in leaky, single-room mud structures.

February 2011 saw heavy winter rains that flooded most tents and left deep puddles and mud everywhere. Night temperatures regularly slid below freezing. There are no phones, no video games, no morale, welfare and recreation resources, and no Internet (so no Facebook). The only mirror for daily shaving is one that rocketed off an M-ATV that hit an IED. It is shattered, of course, but still works—somewhat.

When I met him, Adams and most of the other Marines at the outpost had not showered in 88 days (not a regular shower, anyway). Despite this, they all appeared as if they were on any base back home, always clean-shaven with sharply creased camouflage utilities. They used ammo boxes filled with dirt as weights for PT. Human waste went straight into bags, which they burned in a trash burn pit inside the perimeter of the base.

T-5 has no chow hall; food comes in the form of MREs, and even the "good" MREs get old quickly. Occasionally, Mobile drops off hamburgers and some fresh vegetables like asparagus. But with no refrigerators, what Marines do not eat within a day they must burn or rats will take care of the leftovers. Cola and root beer are always welcome deliveries. Some of the world's most dire poverty surrounds T-5 and, of course, the Marines face the potential for mortar, rocket, and gunfire from all directions day and night. It is a more forlorn, inhospitable to many, nightmarish place than most could ever imagine.

People who have never been to a Marine outpost such as T-5 might imagine the dispositions of those who call it home to be dour and enervated. To the contrary, for the Marines, T-5 represented austerity at its most impressive. There was a lot of contagious laughter and jokes when I was there. They had a ping-pong table set up in an empty mud bunker for the few moments of downtime. Every day at T-5, 2d Battalion, 3d Marines, proves that the Marine Corps truly not just functions, but belongs in every clime and place. "We didn't come halfway around the world to take weekends off," Corporal Dustin L. Ivers, one of the squad's fireteam leaders, proclaimed as he cleaned his M4 [assault] rifle.

Although just 22 [years old], this is the third deployment for Corporal Adams. He had been to Afghanistan with 2d Battalion, 3d Marines, once before and before that was with the battalion for its last Iraq rotation. During his previous Afghan deployment, Adams experienced IEDs on an almost daily basis, including one that struck a humvee carrying two of his close friends. The blast contorted the vehicle so the doors could not open. He watched his friends burn to death.

He is matter-of-fact about such experiences and clearly, despite the popular notion that such experiences and memories precipitate dark, emotional degradation, it imbues Adams, or "Bert" as the other Marines call him, with a level of ardor for each mission that he can virtually guarantee they will find something, help someone, or both. Of course, Adams is not alone in these experiences, nor is his spirit insular. All of the Marines of the squad share his determination.

While Adams and the other Marines of 2d Battalion, 3d Marines, have honed their instincts for seeking out weapons caches, the squad leader pointed to one individual as the thrust for all of their finds: Lance Corporal Parker Hobbs. Hobbs, a Marine Reserve combat engineer from 4th Combat

Engineer Battalion, is the Marine who zeros in on the exact location, down to the inch, of caches of rockets, grenades, rounds, guns, and explosives. He has found about every type of conventional warfare implement.

The specific tactics he employs and his tools, while intriguing, are hardly worth noting in light of his dogged perseverance. While 3d Squad spends roughly equal days patrolling and standing post at T-5, Hobbs ventures outside the wire every single day. When 3d Squad stands post, he is out with another squad. Anyone who has been on a Marine patrol will attest to just how draining they prove to be. For me, joining Marines on a single patrol brings enough physical exhaustion and raw terror to last a lifetime—even when nothing happens. Hobbs does this virtually every day. During one of the rare days he did not venture outside the wire, he told me that he felt not relief of having a day of relative safety, but guilt.

The squad's patrols typically last between 6 and 14 hours and sometimes longer than a full day when they discover large caches of weapons and have to work with disposal technicians. A prime aspect of their mission profiles has the Marines interacting with locals, often keeping Hospitalman Alber Yovani Amaya-Rivera, their corpsman, busy checking on the medical well-being of some Afghans who have never before seen medical specialists.

The going is never easy on these patrols, and they are never typical. Sometimes they move through grassy fields, sometimes they are jumping earthen walls, and many times they have to leap canals, one of the leading causes of injury for Marines in that part of Afghanistan.

They scan constantly for signs of weapons caches in the medium and far distance and always keep an eye closer still for hidden pressure plates or other IED-triggering devices, as well as for potential ambushes. It seems an almost unbearable psychological pressure cooker, thinking that each step, each turn of a corner, could bring a tremendous concussive blast, a wall of black, then ultimately amputation, permanent disfiguration, or death.

No photograph ever could convey this grinding reality. I do not think that any assortment of words could adequately either. Marine Corps ethos dictates mission fidelity above virtually all else, and these Marines of 2d Battalion, 3d Marines, passed outside T-5's wire day in and day out, without fail and without question. If an IED is triggered or a rocket-propelled grenade launched at them and Marines lose limbs or life, the very next day they will be at it again, and again after that and then again, month after month.

And the day after such an attack, if a local Afghan, even one sympathetic to the Taliban, needs aid, these Marines will do all they can to supply that aid. This unglamorous, virtually unseen toil is their purpose. This day-in, day-out toil may not seem herculean, but these Marines are herculean in their resolve and mission fidelity, and they are one squad among many throughout the Marine Corps who share this resolve. Ask any of them in private, and each will tell you that they are nothing special among Marines.

Patrolling and all that it entails always seems to engender unexpected thoughts and curiosities. Toward the end of one patrol where Hobbs, Adams, and his leathernecks plucked a number of munitions from the landscape, I could not help but muse about a subject I do not typically care much to ponder: heroics and the definition of a "hero." It is a term uttered frequently, but not often with regard to the "plain vanilla," or "everyday," patrols run throughout Afghanistan, the results of which almost never are explained to the general public back in the States.

We passed back inside the safety of Typhoon-5's

earthen walls just as a "walk-in" local Afghan source approached to give some information to a Marine charged with intel gathering. "The Taliban are frustrated so much by the Marines here for finding all of their weapons. They know about them all the way back in Pakistan. . . . They can't make a single bomb! They are ready to give this place up!"

Yes, heroes.

Note

Leatherneck 94, no. 7 (July 2011): 32–35. Reprinted with permission. Copyright *Leatherneck*.

About the Author

Ed Darack is the author of *Victory Point: Operations Red Wings and Whalers—The Marine Corps Battle for Freedom in Afghanistan* (2010) about two historic Marine Corps operations in Afghanistan in 2005 (undertaken by 2d Battalion, 3d Marines), reviewed in the August 2009 *Leatherneck*.

Weeks after this embed was over in March 2011, Darack learned that the day after he left Typhoon-5 the Marines of 3d Squad, 1st Platoon, Fox Company, 2d Battalion, 3d Marines, unearthed their largest cache yet—more than 200 107mm rockets—larger than the rest of the battalion combined to date.

Focus on their needs, not what we think they need.
Photo by LCpl Benjamin Crilly

Counterinsurgency as Cargo Cult

by First Lieutenant Matthew F. Cancian

Marine Corps Gazette, January 2013

We Don't Get It

Everything seemed ready for the *shura* (meeting). The cammie net was set up to provide shade, checkpoints were established on the main dirt road, and tea had been prepared. Even with the preparations, however, attendance would probably be low. Despite millions of dollars in infrastructure projects, this area of Sangin, Afghanistan, typically had only a dozen locals show up to the bimonthly shuras where they could meet with representatives of the International Security Assistance Force (ISAF) and one member of the nascent district council. Even bringing the district governor had not increased attendance, and that was disturbing—could he not influence tribesmen two kilometers away from his own compound? At every shura, the locals complained that U.S. overnight patrols were offensive to their customs and religion. Once, more than 100 locals attended to denounce the practice. The human exploitation team later reported that the Taliban had organized their own shura beforehand to instruct people to make that complaint at ours.

This particular shura of ours, however, bucked the trend. A new member of the Afghan interim district council came and showed genuine enthusiasm for his work, controlling the discussion and talking about practical things the government could do for the people, such as registering land, getting identification cards, and settling minor legal disputes. These services resounded with the locals in a way that spending American dollars had not. Was this the beginning of effective local governance in this Wild West? Not in American eyes. Walking away from the meeting, a Marine officer commented, "Well, that was boring. Another shura down, though."

"Counterinsurgency" or Cargo Cult?

Just as Army General David [H.] Petraeus and Field Manual 3-24 *Counterinsurgency* directed, Marine battalions had been conducting shuras in Sangin since they had moved into the district in 2010.[1] But, if shuras were being executed as doctrine directed, why was Sangin still littered with improvised explosive devices (IEDs)? Why were the inhabitants wary, and the Taliban still in control of the population?

Counterinsurgency is a concept that, to paraphrase Sir Henry Campbell-Bannerman, "deludes the earnest and it imposes on the simple." Despite long experience with counterinsurgency, the Marine Corps is still focused on conventional operations, and its application of counterinsurgency doctrine resembles a contemporary "cargo cult." Cargo cults arose in New Guinea in the wake of World War II after tribesmen watched American GIs talk into radios, and then marveled that trucks, airplanes, and supplies arrived as if by magic. Years after the Americans had departed, optimistic tribesmen would talk into long disconnected microphones and build runways hoping that the bounty of the past would reappear. If they ran through the process just right, they thought, good results would follow.

Our application of counterinsurgency doctrine in Afghanistan is similar. We go through the motions of counterinsurgency without focusing on what really matters. This leads to a focus on process metrics instead of outcome—shuras held versus local goodwill, number of partnered operations rather than real relationships built outside the wire, dollars spent versus actual popular commitment, IEDs found versus demonstrated local security forces' readiness. The result is an outlook that is ill-suited to deal with the nuances of fighting small wars, which depend more on understanding local culture and character than on executing time-tested battle drills.

This focus on process manifests itself in many ways. We count and report dismounted patrols no matter whether they have any effect or not, and we count numbers of partnered patrols instead of actually training—and realistically evaluating—the readiness of the Afghan National Security Force (ANSF). We count and report dollars we spend on infrastructure projects instead of looking at what they achieve. Examples of how problems of this nature are preventing us from achieving the outcome we desire in Afghanistan will occur to any observant Marine. Here are some observations from my own experiences.

Counting Patrols

It is another evening in Sangin and that means another overnighter. The Marines patrol to a randomly selected family compound with an interpreter and a small contingent of ANSF. As expected, the Afghan household head is angry and escorts his women to a neighboring compound. In a society where even the slightest insinuation of sexual impropriety can result in death, it is a reasonable precaution. None of the overnight patrols in this area have encountered an IED emplacement (much less actual Taliban fighters), so why do the Marines continue to perform these intrusive patrols? That the population is the battlefield in a counterinsurgency is repeated endlessly in all the literature. And so the company order reads that the platoons will "conduct 24-hour operations in order to secure the local population." But it is a lot easier to measure the durations of patrols than the effect they have on the population. The consistently negative response displayed by the locals gives a good hint as to the latter, if anyone wished to see. No matter. We patrol in order to be able to report hours spent patrolling.

The same fixation on the metrics of patrolling bedevils our attempts to improve the readiness of the ANSF. A major part of this effort consists of "partnered patrols," in which Afghan soldiers are supposed to accompany U.S. forces to learn our methods and, it is hoped, to be inspired to emulate our professionalism. But there is no attempt to judge what, or how much, the ANSF learn. What is judged, meaning reported to superiors and Marine officers evaluated for, is the simple number of such partnered patrols.

Commanders, rationally working the metric, try to conduct as many such patrols as they can regardless of the quality of ANSF participation. An Afghan fireteam will accompany a Marine squad, or even a single Afghan policeman will be dragged along, chosen because he was the least high on hashish that day. The daily tragicomedy of partnered patrols as they often work in reality is familiar to every Afghanistan Marine, and has even made its way into the newspapers.[2] Positive interactions are possible. In a different area of the Sangin District, the Marines patrol to a local's compound in the daytime and find themselves playing the board game Chaka Chaka Punj with the inhabitants. While the Marines have never played the game before, its similarities to Sony soon become apparent. An officer shares his discovery with the Afghans that, while Christians believe that Isaac was almost sacrificed by Abraham, the Afghans believe it was Ishmael. The Afghans are surprised not at the difference, but at the similarities. A discussion ensues about their attitudes toward IED strikes on locals—a discussion, rather than a lecture, by the Americans. A few days later, the locals give the Americans a puppy for their base.

Money as Metric

Of the $900,000 worth of projects being run in the Sangin area, the civil affairs officer was particularly proud of the school. And why should not he be? The next two largest projects were roads between American patrol bases for American military use. While the Afghans were quite happy with winding dirt roads, the U.S. maneuver commanders wanted straighter lines of communications (regardless of the occupied Afghan compounds between them). The concrete structure of the school will certainly stick out amongst the mud brick houses of the Pashtuns, although not for its size, as it will only have four small classrooms in it. But after four months of work, the provincial education official tells the Americans that they must spend another $100,000 to install electricity in the school or else the Afghans will not provide a teacher. *Insha Allah* (God willing), no other "requirements" would be discovered before the project was completed. Nevertheless, the project is deemed a great success because success is defined in terms of dollars spent. As Bing West points out repeatedly in *The Wrong War*, however, unfocused development funds do not buy security. "How we spend is often more important than how much we spend."[3]

The disconnect between American dollars and Third World reaction is hardly new. Indeed, the theme of William J. Lederer and Eugene Burdick's 1958 classic novel *The Ugly American* was that American development efforts in Southeast Asia were grandiose, but useless to the locals. In

Photo by Cpl Skyler Tooker

Are we going through the motions or truly focused on outcomes?

the book, Americans who truly make a difference were those who focused on small, incremental improvements in the everyday life of the natives, such as introducing bicycle-powered water pumps and dairy livestock.

There is rich literature about what makes development effective—that being a focus on recipient needs instead of donor wishes. The Marine Corps appears to ignore this. For example, in some areas of Helmand Province, small landowners are reluctant to grow wheat because, in order to get it threshed and milled, they have to pay as much as a quarter of their harvest to rich landowners to use their mills, which cost around 200,000 Afghani (USD $4,550). It was proposed to buy a few of these machines and put them at the government center for the people to use free or at a much-reduced price. Such a service would also bind farmers to the government. A $15,000 bill for three of them would have been a pittance in this context of Marine development expenditure. The project, however, was stillborn because it was not spectacular enough in dollar amount to grab attention.

The Way Forward?

Our current approach to many problems in Afghanistan focuses on form rather than substance, processes rather than results, and penalizes efforts that focus on substance and results. Any Marine can give you examples, and here is mine. First Lieutenant Smith is a highly motivated leader of a training team. He focuses on developing close ties with his ANSF partners, and they respond by beginning to conduct independent patrols of the area, becoming steadily more proficient and disciplined soldiers. Due to manpower constraints, however, he has a corpsman drive a vehicle in order to visit his far-flung posts. First Lieutenant Jones, on the other hand, believes that working with the Afghans is a waste of time and that all that "nonkinetic" stuff should be dealt with by someone else. He does, however, report a large number of partnered patrols by dragging a handful of resentful ANSF along with each of his Marine patrols. You know how these stories end. At the end of seven months, Jones leaves his area of operations completely unchanged, but has put up impressive numbers and receives an excellent fitness report. Smith was quietly relieved of duty when higher headquarters find out about his violation of the rules concerning the use of corpsmen. Is it surprising that we are breeding a force that values gaming the metrics over accomplishment of the mission?

But how do you reorient a whole military culture to substance and results? The first step is to recognize that rule-based management of counterinsurgency can be counterproductive. Lieutenant Colonel Douglas J. MacIntyre wrote an excellent article calling attention to the problem of making good metrics, and had several suggestions that, while abstract, are based on good principles: bottom-up assessment, tactical learning, and orientation on the desired endstate.[4]

Instead of blindly pursuing metrics that look good for the "commanders update brief" while secretly hoping that we get to pulverize some Talibs, we need to understand the underlying dynamics of counterinsurgency. Few officers have read Field Manual 3-24, much less David Galula, David Kilcullen, histories of Afghanistan, or accounts of Afghan culture. How then can they be expected to win an unconventional fight? Units need to integrate education on counterinsurgency into their predeployment training to fill this void. Once maneuver commanders are more aware of local conditions and the lessons learned about other counterinsurgency efforts—lessons that have been written in blood—they will adapt themselves to the demands of the fight and make better decisions, leading to greater success. Mindlessly tallying up partnered patrols and the dollars we spend, then expecting success in counterinsurgency, is the equivalent of building bamboo air control towers and looking hopefully up at the sky.

Note

Marine Corps Gazette 97, no. 1 (January 2013): 50–52. Reprinted with permission. Copyright *Marine Corps Gazette.*

About the Author

First Lieutenant Matthew Cancian deployed as a forward observer with 1st Battalion, 5th Marines, to Sangin from March to October 2011. He was the executive officer, Headquarters Battery, 1st Battalion, 11th Marines.

Endnotes

1. Gen David H. Petraeus, USA, "COMISAF's Counterinsurgency Guidance" (memorandum, Headquarters, ISAF-Afghanistan, August 2010).
2. Matt Millham, "Afghans Show Little Progress as Security Hand Off Draws Near," *Stars and Stripes*, 8 August 2011.
3. Bing West, *The Wrong War: Grit, Strategy, and the Way Out of Afghanistan* (New York: Random House, 2011), 273.
4. Douglas J. MacIntyre, "Operational Assessment: Replacing Black Magic with Learning by Design," *Marine Corps Gazette* 95, no. 6 (June 2011): 17–22.

The Marines with 1st Battalion, 6th Marines, conduct a high rate of patrols in an attempt to limit the insurgency's freedom of movement in this key district.

Defense Imagery 491110, photo by Cpl James Clark

Interview: First Lieutenant Dennis A. Graziosi

by Marine Corps History Division

Marine Corps field historian Lieutenant Colonel Mark Wood interviewed First Lieutenant Dennis A. Graziosi, 2d Platoon commander, B Company 1st Battalion, 6th Marines, on 12 February 2012 concerning Graziosi's 2011 deployment to Afghanistan. The following has been edited from the original transcript for space constraints in this work.

Wood: So, when did you go to Afghanistan?

Graziosi: We left in July of 2011. It was 14 July.

Wood: And, where did you go to?

Graziosi: We flew into Sangin [District]. We were patrolling, I believe, six days after we left and we spent the first couple months of deployment in Sangin patrolling the green zone. Originally, we had three positions. And then, right before we left, we turned over one of the positions to our platoon and two positions before—we turned over to 3d Battalion, 7th Marines, in preparation for Operation Eastern Storm.

Wood: Okay. Can you describe— When you got to Sangin, what was the operating environment like?

Graziosi: Initially, it was a huge difference from what we were used to. Everything we had trained for was pretty much thrown out the window. We trained for pulling formations, contract drills, trying to work the counterinsurgency piece while possibly taking gunfire. You go out there and everything's completely different. It was patrolling in a straight line, in Ranger file. Sweeping for IEDs; visual confirmation was almost never there. Marines may [have] had to have used some kind of probing techniques to find an IED after they got a metallic hit. It wasn't kinetic as far as firefights goes, but it was extremely kinetic as far as the danger in the area; finding IEDs, pressure plates, smaller explosives with pressure plates usually causing single or double amp. They were just— The enemy was emplacing them wherever they thought we would travel or they thought the people would travel to get to us. So, it was a lot different than we had expected. We hadn't heard much about it before we got there. We just knew there was a lot of IEDs there, but we didn't know we're going to Sangin until right before we left. We adjusted quickly to the area we were at.

Wood: So, where in Sangin were you?

Graziosi: We were— My positions are right along the Helmand River in Sangin in the green zone.* It stretched just north of the company position from the Helmand River the whole way over to Route 611, the main MSR [main supply route] there in Sangin. They were pretty much all spread along the river, trying to provide a blocking from coming over across the river into the green zone.

Wood: So, you said there was a lot of IED activity. So what kind of things— You said there were strikes and wounded Marines. Was that a pretty frequent occurrence?

Official U.S. Marine Corps photo

U.S. Marine Corps 1stLt Dennis A. Graziosi.

Graziosi: No, not so much. We had a lot of IED finds. Just in the Marines doing good TTPs [tactics, techniques, and procedures]. We could have had a lot more casualties than we did. We actually only sustained one IED strike while we were there with my platoon. There was another one right as we were leaving with the new unit that was coming in. Those were the only two IED strikes that we had while we were there. However, I think we found over 40 IEDs in the area. And some of them were old, and I'd say about half of them were new. We were basically finding old ones that— The British had come in 2006 and they pretty much blew up everything. They saw— established PVs and did patrol. And if they did patrol it wasn't that far, so they— Taliban would place IEDs as kind of like a defense and as a result we kept finding a bunch of old IEDs and about half of them were new.

* The *green zone*, in this instance, refers to the agricultural area that runs adjacent to the Helmand River.

Wood: What do you attribute to being able to find most of these, rather than them going off?

Graziosi: I would have to say at the end of the day it was good TTPs by the Marines. We had the technology, you know, minesweepers, the THOR systems.* It didn't really help all the time. When it came down to it, it was the squad leaders planning their routes and knowing where they would put an IED if they were the enemy. Then, down to the actual individual point man who's sweeping, waiting to hear a metallic hit, taking things slow. Had we have rushed or gone into areas we didn't need to go or taken obvious routes every single day, I think, maybe it would've been a lot worse. But as a result, we were able to find taking it slow and not setting patterns, I think, that's without a doubt. When it came down to it, it was the individual Marine just doing his job very well.

Wood: What was the COIN [counterinsurgency] environment like?

Graziosi: In Sangin, it was unique in that the British had been there in 2006 and there were several battalions there before. So, people knew what to expect from us. They knew— They saw Americans as powerbrokers that can pass out money. Consistently wanting things from us. We immediately stopped that and just worked on building a rapport with them. The counterinsurgency there—for our area, worked a lot better in Sangin just because they were used to us being there. There was a village just to the south of my position called the Brockside Village. And prior to us getting there, it'd been a village that was about a hundred percent with Coalition support and kind of playing both sides of the fence. But just through our increased patrol efforts and spending time with them, that ultimately led to our success there. A lot of times us finding the IEDs, the locals would point them out to us. And that didn't happen until after we established a relationship with them and then anytime an IED was put in, they would instantly tell us. They may not know exactly where it was, but they would say, "Hey, the Taliban was here last night. They were in this general area. There's probably an IED there." At first, we relied solely on our tactics and then, by the time we left, it was if someone came in and talked to us, we knew there was something in the area. If nobody came and talked to us, we felt a little more confident in the area.

Wood: What was your focus? Were you focusing on the COIN engagement or the tactical side, or—

Graziosi: Me personally?

Wood: Yeah. Yeah.

Graziosi: Me personally, it was tasking my squad leaders with the right kind of mission. I didn't want them to focus solely on the people and forget that there was an enemy out there. I wanted them to focus on the whole sphere of the conflict we were in so it would be— Whether it's go out, don't talk to people so much, go establish a blocking position or an over watch position, sit for a couple hours, let me know what you see and then come back. Or, was it go out just to talk to the people, establish a relationship with them. It was— My focus was constantly keeping them focused on a mission; not just walking out every day, exposing themselves for no reason. Trying to find a mission and intent for them and tying it into whatever our goal was. Turn it over to the

* The THOR system is a man-portable, counter-radio-controlled IED jammer.

Afghans and make the area as safe as possible for the next units coming in behind us. So, I would say my biggest focus was trying to give them the right missions at the right times, focusing on the right things.

Wood: How were the Marines? How was the morale? How did they feel about what they were doing in the Sangin [District]?

Graziosi: That's an interesting question because when it comes to morale, Marines aren't complaining about something, they're not—doing their jobs. They didn't like the COIN piece. Marines are trained to go in and blow everything up, shoot everything and kill the bad guy. This is more of a hunt and kill mission. It wasn't like, "There's the enemy go get him." It was work with the people—it was almost like— An average Marine was— To be a Marine you had to do the regular Marine things. You also had to be a policeman. You also had to be a politician, to work with the people and be able to talk to them. Twenty-one-year-olds holding shuras with 15 plus people; it was almost like they had to put the face on of being like many different things at once. So, it was very difficult for them. They did outstanding. And at the end of the day, their morale to go out—go out of the wire every day was never a question about it because if someone when their squad was going out, everybody was going out. But the morale was a little bit different because they want to go out and do the things they were trained to do; not go out and win the support of the local populace. It's a little difficult to stomach for everybody.

Wood: Okay. So, you spent about two months in Sangin. Can you talk about when you found out you were going to be going to Kajaki [southern Afghanistan], correct? You know, that whole process.

Graziosi: We had been hearing rumors ever since we left that we were only going to be in Sangin for a month or two. And from there, we were going to pull out and do a major mission in Kajaki. And, for me, that was almost cancerous to our unit; I viewed it as [such] because I didn't want them to think that we were just marking time before we did our real deployment. I never thought we were going to do Eastern Storm. I just thought it was too late in the war to do a major operation like that. So, I tried to keep everybody focused on— Our whole deployment was going to be seven months in Sangin. That's what I went with. That's what I wanted my Marines to go with just so they would stay focused on the task at hand. And then, I would say about a week and a half before we actually pulled out of Sangin, it was a hundred percent final. We had heard it was a go, then we heard it was a no go. And then it was a possibility that another unit was going to do it not us. So, it was constantly going back and forth. But then, once we heard 3d Battalion, 7th Marines, was showing up to relieve us— There was a little bit of confusion between how the battalion that we were attached to—1st Battalion, 5th Marines—was going to turn over the battlespace to a lot less Marines than they had in the area. Obviously, we were a company attached to a battalion, so we had a lot of extra presence there. As we pulled out and 1st Battalion, 5th Marines, was turning over, it was just kind of difficult how they were going to manage with the less Marines in the area. Once they actually started arriving that's when we knew it was a hundred percent.

Wood: So, can you talk about the pulling out and what the plan was? What the objectives were?

Graziosi: For us, it was trying to give them the best turnover that we could. We had marked down all of the significant acts that had happened. Everything that was pertinent to the area. We had taken pictures of all of the locals. Where they lived, what house they were in, all that kind of stuff. We just tried to provide them with as much information— In Sangin, too, we did have a good ANA [Afghan National Army] source with us. A lot of our Marines established good relationships with them. We patrolled a lot with them so a turnover with them was pretty relevant. Trying to get them to do as much of the work as possible and try and set them up to do the dangerous things that we were doing. Sweeping and walking the point on patrol; trying to set them up for success. Trying to give them as much information that we had gained as possible.

Wood: So, it was a pretty quick turnaround from finding out you are actually going to go to Kajaki.

Graziosi: Yeah. I mean, you technically could have said that we knew about it the whole time, but we just didn't know it was a hundred percent until about a week and a half out.

Wood: So, what actually happened, from the time you left Sangin [to] the move to Kajaki?

Graziosi: We'd left Sangin. All of our forward positions had pulled back to the company position. A lot more secure area, showers and everything like that. I think we had about five days for planning; detailed planning, rehearsals, order issue, and prep for execution, which wasn't a lot of time at all, and we had just finished doing the relief. So, it was a lot of things going on at once. And the rest of the time was just spent trying to find as much out about Kajaki as possible in a couple days, getting ready to do the actual mission.

Wood: Okay. So, what did it— So, your platoon— What was your platoon's mission?

Graziosi: My platoon was the main effort. Our mission was to clear the entire green zone and establish a patrol base behind the bazaar of where population center was and establish a patrol base there. We landed at the same time as their platoons. Stayed out for about five days, and our mission was just disruption, patrolling throughout the green zone. It was technically a clearing mission, but you can't really clear that area. As soon as you do, if you backfill it instantly in patrol bases—we covered too big of an area to actually backfill; it was just a major disruption op which proved beneficial for us.

Wood: What were some of the highlights of that?

Graziosi: It wasn't as kinetic as we thought it was going to be. They were definitely prepared for us. There were a lot of bunkers. We were doing a traditional assault from the roads. We would've taken a lot more casualties. Even with our scheme of maneuver, a lot of moving parts and a lot of deception that was involved with it. They still could have inflicted a lot of casualties on us. I would say that our success the first 24 hours companywide was what made them ultimately leave. The first thing that happened to us, we had moved about 500 or 600 meters through open area and I was extremely nervous because there were no compounds for cover. And we just did squad bounding over watch. I had four squads in my platoon. We just kept bounding squad by squad. Tried to give the Marines some kind of rest with heavy packs. We had gotten to the first set of compounds and set one squad up and I pushed up with our fourth squad. We had saw a bombed out building. The first two Marines up front observed two guys with an RPG; was trying to run away and they engaged both of them. And I immediately start[ed] rendering aid to them. We had gotten their ICOM radios and pulled my interpreter up with me and I was listening to them talk and try to get radio checks with the two that we had shot. They said, "They're not there. They're not there. Where are the Marines at? I don't know. We can't find them. We think they're coming close. Where are those other two guys at? I can't remember their names or what their call signs were." They said, "Go there. Bring the machine guns. Prepare for the attack."

At that point, I knew that my platoon was exposed. We were still trying to move up. I had two squads that were still back in the open area and then I had one squad forward. And then, one squad was actually in one of the compounds. I pulled my whole platoon inside the compound—the two compounds just so we could at least be in and cover in case there was actually an attack, which there ultimately turned out to be. Once we had consulted ADEC [aviation data exploitation capability], there was one squad forward with eyes on both of the enemy and eventually one was killed instantly. The other one was alive until the attack happened; he had died somewhere in the process. Instantly we had— As soon as we got everybody inside of the building— I remember it almost— Once everybody was inside was when it all just started. Set out small arms and then pick up machine-gun fire. It was kind of difficult to tell who exactly was shooting; it was just extremely loud. Then, one of my squad leaders had a position on top of the roof, took some accurate rounds—rounds had packaged in front of his face. He was kind of dazed for a second. I saw his legs go limp. I thought he got shot. We pulled him down instantly. Realized he was okay, and at that time we had a [M]240 [machine] gunner picked up and was ready to fire. And then, as he picked up, it was shooting pretty fast. Almost between the rapid and cyclic break, I had eventually put them off. From there, there was a couple of sporadic fire here and there but that was part of us taking out a certain number of guys, however many it may be. Definitely two killed in action. Probably four to six more wounded. That mixed in with the company's success with all of the platoons. [The] 1st platoon was in a major firefight; 3d platoon had taken some contact up to the north. Just our instant success within the first couple of hours, I think that just pushed them off. They didn't want to stand up, they knew they couldn't fight us face-to-face. They pulled back across the river and waited to cross back again and engage us.

From there, we had done patrolling disruption ops in the green zone. We had found three pretty substantial IED cache sites with about 200 pounds of HME [homemade explosives] total. And over all, it was 91 pressure plates, components, some manuals, some pictures, and a bunch of other stuff. But after that, we had pretty much taken out their supply a lot. We still stayed out for about five days and ran out of food and water a couple of times. Got some resupplies and then ultimately ended up on Patrol Base Pennsylvania behind the Kajaki Sofla Bazaar.

Wood: Okay. And does that kind of take you through— Well, what happened following that through the rest of the deployment?

Graziosi: After that we just did stability operations; hitting counterinsurgency. There was a little bit of a lull for a while. Basically, the people were starting to move back in. We were quick to get out there and patrol, and engage and meet them. Trying to work with them. What we soon found out, based on the intelligence and based on everything that had happened, everybody in that area had some kind of ties with the Taliban. It was just a huge safe haven area. The people either chose to support them, against their will supported them, or however the case may be. They were all in some way, shape or form, associated with the Taliban. And so, when the Taliban came back to hit us they had some safe house to always stay at. And from there, their TTPs turned into trying to front lay a DFC [direct focused charge] IED or try to emplace one real quick.

We had one IED strike outside of a bunker. We tried to call an air strike in the bunker but there was compounds too close to it. One of my Marines tried going into the bunker and they had emplaced IEDs outside of them. I think we had blown up 15 bunkers at this point and the ones out by the river they had placed IEDs in them. And one of my Marines stepped on an IED and lost a leg. We evac'd [evacuated] him out of there. A couple— I'd say about two weeks later, another Marine was walking down an area where we had known the people; we had seen them, talked to them. They were all around the area; Taliban just happened to slip a DFC IED. While he was crossing the bridge, caught some shrapnel. Urgent casualty evac'd him out of there. Two days later from there, same thing. We saw, on 10 November, two of my Marines were attempting to cross a canal. At this point, we had stopped using bridges because we knew they were trying to use them for an IED spot. They had front laid an IED. Luckily, the Marines had good dispersion, good standoff. They just caught some shrapnel. They were okay. Still urgent casualties. We evac'd them out of there, but at the end of the day it was— We're lucky it was just shrapnel and not somebody getting killed trying to cross the bridge. After that, we had done a major cordon and search on a village. I had pushed my whole platoon out. Requested support from air platoons to come establish walking positions, and we just basically showed up [at] three o'clock in the morning, sat and waited until morning prayer, when everybody came outside. The Marines were all spread out enough so we could see, cover a lot of the area. I think we had detained about 10 people that we had intelligence on, one of them being a JPEL and a higher-level Taliban commander.*

We had done several night raids like that that had led to us detaining two Taliban commanders and killing one in [an] engagement across the river. The only firefight that we had got in after the clear, we had observed what they're—suspicious individuals moving back and forth between the tree line where a bunch of bunkers were. We also heard ICOM traffic saying, "We can't put the gun here; the Marines have blown up the bunkers. Where should we put them?" So, we watched a little bit closely and, simultaneously, I sent a squad up. And they had observed— We observed them with their—run inside a building. Corpsmen searched that building and found some IED-making materials. A small cache, compared to what we had found. But it was definitely something that was actively being used, which was why there was a lot smaller amounts. It wasn't really storage. It was "get-in-and-get-out" usage area. All my squad was— One of my squads was out. Close to the river, they observed two of the Taliban with weapons across [the] river. And we had tied together the screws the team that was trying to bring the machine gun in our area. And [we] suppressed them—make the suppression with

* JPEL, or Joint Prioritized Effects List, is the list of individuals targeted by Coalition forces.

machine guns, called in airstrikes, and got the ICOM traffic afterward that the Taliban commander was killed. Never could really confirm that, but from what the people said, it was a Taliban commander and about two other guys were killed in the airstrikes. After that and a couple cordon-and-search missions at night and some of our other success that we gained working with the people. Everything was quiet for the last month and a half, two months, and there was just nothing the Taliban could do at this point. The people were with us; they supported us. We had found all of their supply routes. Weapons Company had pushed across the river, which put some pressure on the Taliban, which ultimately led to the suicide bomb that had happened at the bazaar a week before my Marines were supposed to leave.

Wood: And that was— Where was that bazaar?

Graziosi: Kajaki Sofla Bazaar.

Wood: What happened with that?

Graziosi: Basically, we had gotten intelligence that there was a suicide bomber that was trying to come into our area and hit one of the ADP [Alternative Development Program] Afghan police stations that was close to our base. While one of my squads was coming back from patrol they were crossing Route 611, which is the main road. It also separated both sides of the bazaar, and a suicide bomber had been waiting. He drove around, came in through a different entryway to the bazaar, and blew himself up, trying to just kill

Defense Imagery 484183, photo by Cpl James Clark

A Marine with 2d Platoon, Bravo Company, 1st Battalion, 6th Marines, provides security during a patrol through the Kajaki Sofla Bazaar on 26 October 2011.

as many people as he could. Two AUP [Afghan Uniformed Police], at the last second, tried jumping on top of him. Those two were killed instantly. One of my Marines was killed, another severely wounded, and two others urgent casualties that we had evac'd. There were countless wounded or killed that we could see, and we never got a solid number of how many. We evac'd 27 people out of my patrol base. I can't remember the total statistics; I think there were about 10 killed and 20 plus wounded that were evac'd either at my position or at the battalion position, FOB White House.

Wood: Locals or—

Graziosi: Locals. I'm trying to think here for the statistics. I think two AUP were killed, the two that were closest. I think there was about 10–12 local nationals that were killed. About 20 additional that were wounded, all local nationals. I had one Marine killed-in-action; one wounded severely; two other urgent casualties that had taken shrapnel or burns. Four casualties total with my platoon. We just started coordinating evac. I was at another platoon position when the actual explosion happened and the vehicle was getting ready to come back. I was doing an accountability piece . . . some of the weapons, doing a turnover with the new lieutenant. I had seen the smoke cloud and we'd gotten the [victims] and instantly got calm with company. I remember asking our company, "Hey, what's going on? Was that an explosion?" And, "We have a mass CAS [close air support] at the bazaar. We're trying to get calm with 2/2," which is my 2d Squad. And I remember talking to my squad leader the night before, and I was saying, "Hey, the last couple of days we've been out on patrol for four to five hours. Being out there a little bit longer than we expected. Let's make this quick. Try to change it up a little bit so we're not setting a pattern of being exposed for that long." I looked at my watch and it was exactly two hours after they left, so I knew something was bad. It was all with them. When I got to— I hopped in the vehicle real quick. We were there within minutes. They were still rendering aid when I got there. Just looked at what was going on; there was just people everywhere. Body parts everywhere. It was pretty gruesome. Pretty bad.

Wood: Did you get a sense for what the locals thought about that? What I'm trying to get at is, did the Taliban get the intended effect?

Graziosi: Yeah. I think their intent was to hit the police, as well as us, as well as some of the locals that were talking with us. What I thought was going to— I think what the Taliban had anticipated on happening was the people were going to see that they had the power to do this. They were going to back off. They were going to just be neutral in hopes that, whenever we did leave, they could come back in and retake their area. They had wanted the people, not so much to support the Taliban; they knew that wasn't going to happen with a suicide bomb. They want us to stop being so supportive of us and just be neutral. Back off to the point where they're not working with us, they're not talking to us, but they're not hating us. They're not working against us. It couldn't have backfired worse for them. Unfortunately, when there's a local national casualty, that's the best thing for Coalition forces so that they actually see what the Taliban has the capability of doing to them. The very next morning, I held a shura in the bazaar right where it happened with the locals. And we just sat down and talked, and every single one of them were like, "No we're working with you guys." Because I don't think the Taliban wanted to have that many local national casualties. I think they wanted it to be just one or two that was talking with us, one or two that was talking with the AUP. And the people pretty much all came together and said, "We're done working with them. We see what can happen." We just basically tried to exploit it to work in our favor. "This is what happens when you help them. This is why you should help us and not help them. They don't care about you." We felt the pain; the sacrifice that we made. We look at the AUP. They had lost several and the local nationals lost the most out of all of us. Just a significant amount more than we did. So, all of us suffered together and it was definitely not what the Taliban had intended.

Wood: You mentioned the success of quickly establishing supremacy in the Kajaki operation. What other successes stand out?

Graziosi: It was half civilian and half militarily, if that makes any kind of sense. Half of our tactical missions cordon insurgents, it just always seemed like we were catching them at the right time. We were— Our intelligence that we would get a lot of the times we don't believe it to be a hundred percent accurate. Anytime we would [go] to a cordon and search on a building, I would always say, "We're out here. We are searching way more than we need to." And on two separate occasions, that helped us out finding some of the guys we were looking for. So, a lot of success operationally with cordon and search and also the counterinsurgency piece. Just really try to exploit the people by showing them that we can establish relationships with them; that we can work with them. We used one of our landing zones inside of our base as a soccer field and the kids would come up and play soccer with us. That helped out; build some rapport. We built a school just like we did in Sangin where— The way I would do it was squad leaders would go out and talk to people, "Hey, we want to build a school. Come back and let's have a shura." They come back, and I'd say, "Hey, you know GIRoA [Government of the Islamic Republic of Afghanistan] is going to build a school here soon. If we start, get ahead of the game, volunteer, they come and see the progress that we've made. Then we can get government support." And people were all about it and they started holding classes. I mean, it wasn't anything organized as to what we know is a school but it was just *mullahs*, or elders, somebody smart getting in some kind of school. Whether it's in a mosque or outside. Just talking to the kids and instantly getting support from battalion. Getting books out and having the kids write in them and bring them back to the base to show us definitely helped out. So, our success was half operation and half counterinsurgency with the people.

Wood: What memories stand out to you?

Graziosi: For me, all I can think of was how great my Marines performed. I mean, it was just— Everybody says that. Everybody says, "Wait until you've done deployment to see [what] these guys can do." Until you can see it with your own two eyes and you can interact with them every single day; they are just constantly impressing you. A lot of times, as we see now, you're having to work through the hard things with the Marines, but when they're out there and s——t hits the fan, they're really— They know exactly what to do and they'll do anything for their brothers. And all of my memories stand out is with them, joking around with them, watching them joke around, being involved, going on patrol with them, hearing them talk, and just being beside them as much as possible and trying to take care of them as much as possible. I can remember once the suicide bomb happened, by the time I got there, everything had been in place.

They had already been doing the right things. The only thing they needed from me was to say, "How are we getting these guys off the hill? Should we wait for our vehicles to come down?" And I knew if we had done that it would've taken too long. Then we may have lost one or two more and we were stopping police vehicles. As many people as we could fit on these vehicles, I don't care how they get up the hill just get them up the hill. That was all I needed to say and instantly they were already on top of it. Putting people on the back of trucks. Truck after truck getting them up the hill. And then, instantly starting triage, first aid, the medevac process, establishing security in case they were trying to hit one of the birds coming in to pick up any casualties. My memory stands out of just them knowing just exactly what to do. And I would just say their overall attitude. You know they didn't want to do the counterinsurgency. They didn't want to go out and operate with people but they would still do it and they would do it to the best that they could to the best of their abilities. They disagreed with it, they didn't want to do it. No Marine wants to go out and shake hands, they all want to go out and get [in] firefights. But at the end of the day, they're going to get their mission intent, they're going to go out and accomplish it in any way, shape, or form that they can. I'm trying to think of anything else that stands out. I would just have to say that everything they did together and when things got really bad they knew exactly what to do. A little bit of encouragement and a little bit of guidance and they would just take it and run with it.

U.S. Marines patrol near Forward Operating Base Musa Qala, Helmand Province, Afghanistan.

Defense Imagery 1037387, photo by LCpl James Mast

Interview: Corporal Raymundo Mendez Romero

by Marine Corps History Division

Marine Corps field historian Lieutenant Colonel Mark Wood interviewed Corporal Raymundo Mendez Romero, a fireteam leader with 3d Platoon, Echo Company, 2d Battalion, 4th Marines, on 7 June 2012 concerning Mendez Romero's 2011–12 deployment to Afghanistan. The following has been edited from the original transcript for space constraints in this work.

Wood: When did you all arrive in Afghanistan and where was that?

Mendez Romero: We arrived in Afghanistan around September—the first week of September.

Wood: That's good. Yeah, I mean approximate is fine. And where was that location?

Mendez Romero: We arrived to Musa Qala [Helmand Province, Afghanistan].

Wood: Musa Qala. Okay. Can you talk about what your first impressions were when you got there?

Mendez Romero: When I first got there, it was all so fast. It was just one second we were in the company FOB [forward operating base] and two hours later we were in our platoon FOB.

Wood: Okay. What was the— Did you do any

turnover with the previous unit that was there?

Mendez Romero: Yes. When we got there, the platoon that was there, each of their squads took each of our squads. We shadowed their squad and we went out on patrols with them. They showed us which areas were hot, hostile, and then which areas were like secured in a way.

Wood: Mmm-hmm. Okay. Where was this? Where was your platoon [position]?

Mendez Romero: That was Salaam Bazaar.

Wood: Salaam Bazaar. And is that close to . . .

Mendez Romero: It's in between Musa Qala and Now Zad.

Wood: Okay. All right. So you got there, you did the turnover, and what was your basic sense from what was going on? Was the environment permissive? Can you just kind of describe the way it was?

Mendez Romero: When we first got there, it was hot and there were farmers outside. It looked like a normal life for them. And, eventually we started like pushing out the patrol base farther out and that's when we started receiving contact.

Wood: Okay. So, let's just keep moving on from there. So you started pushing out about how long after you'd been at your POS [position]?

Mendez Romero: I personally— It was fireteam leaders and squad leader, they went out to the squad first and we got familiarized with the area first. So, that was like in the first two days.

Wood: Uh-huh.

Official U.S. Marine Corps photo

U.S. Marine Corps Cpl Raymundo Mendez Romero.

Mendez Romero: And then after that it was regularly.

Wood: Okay, regular after the first couple of days. Okay. When you went out, you said you saw the farmers and kind of doing their regular situation, carrying on their life. What were the types of first contact that you had with the enemy?

Mendez Romero: The first contacts we had, we were talking to locals, to an elderly and eventually we . . . arrested him in a way. And then when we were doing the process of arresting him, when [the] QRF [quick reaction force] came out, we started receiving pop shots from 500 meters and out. And then we closed in on the enemy and we started receiving contact from another position.

Wood: What happened from there?

Mendez Romero: We had three teams. One team was on the south side of the area, then another team was on the east side of the compound we were in. And my team, we started taking contact from the north, so we started rushing toward the contact and that's when we received contact from the south. And the south team, they started engaging the contact. They took out the threat and then we— [The] QRF was already there, so they pushed out farther and they tried covering [and] moving with us to provide support.

Wood: Was that your first time in combat actually being shot at?

Mendez Romero: Yes, it was, sir.

Wood: What stands out to you from that experience? What do you remember thinking?

Mendez Romero: At first, I couldn't believe it.

Wood: Yeah.

Mendez Romero: It was just like, not a shock, but it was just like a surprise. Like real life hitting me. And I embraced it, and I kept moving.

Wood: Okay. So that was just in the first week or so?

Mendez Romero: Yes, sir.

Wood: Right. So how did things go from there after that first week?

Mendez Romero: After that first week, we started— We would take turns with the other squads patrolling.

Wood: Mmm-hmm.

Mendez Romero: They would receive contact, sometimes earlier than we did, from like the yardage, which is like 150 meters out. The PB [patrol base], they would start receiving contact. And we would support, and they just kept going like that for a while, and then we eventually closed down that PB.

Wood: Mmm-hmm. What was kind of the— What was the patrol base like? Just describe what was there.

Mendez Romero: The PB was [a] pretty decent shape or size. It was like a triangle; it was pretty much a triangle and it had c-wire [concertina wire] all around it. About 40 yards outside of the PB [were] Hescos [barriers], it was c-wire for like a bigger security. And then we had three posts on each corner of the PB. Well, we had a post on each corner, so we had three posts total.

Wood: Yeah.

Mendez Romero: We had an ECP [entry control point] and we have our COC [combat operations center] and then we had tents for each squad.

Wood: All right. And you had tents and what was— Were you getting a hot shower or anything like that?

Mendez Romero: We had to— There was these bags that you put in water, you'd leave it in the sun, and the bags would heat it up on its own, like solar power in a way. And after that you would take a shower. That's if you had time.

Wood: Okay. If you had time. What was kind of the— During these first few weeks, what was the schedule like?

Defense Imagery 273277, photo by Cpl Daniel Blatter

Marines roll through the Salaam Bazaar in Helmand Province. Enemy contact was imminent as they patrolled through the known Taliban stronghold.

Mendez Romero: The schedule was like one would be a squad on patrol and a squad on post. And then weapons, the weapons detachments, they would rotate in and out with us according to what we were scheduled to.

Wood: What was your sleep plan like?

Mendez Romero: It depends, sir.

Wood: Yeah.

Mendez Romero: It just depended on your day of return.

Wood: Were you all going out at night too?

Mendez Romero: Yes, we did night patrols as well.

Wood: So going out there, you said you had contact. Any particular— You mentioned the one incident when you brought in the elder. Any other incidents stand out to you from that period?

Mendez Romero: From that period, no, sir.

Wood: Okay. So you closed down the patrol base. You'd been there for a while. How long had you been there?

Mendez Romero: I would say a month and a half.

Wood: Okay. About a month and a half there. What happened from then, after you closed the patrol base?

Mendez Romero: We closed PB Salaam Bazaar

and then we closed another PB at the same time.

Wood: Uh-huh.

Mendez Romero: And then we moved back to the company FOB.

Wood: Where was that at?

Mendez Romero: That was located in Musa Qala.

Wood: Okay. So what was your platoon doing after you moved back to Musa Qala?

Mendez Romero: As soon as we got back to Musa Qala, because our squad was in charge of making sure that the PBs were completely like flattened out.

Wood: Like demilled [demilitarized]?

Mendez Romero: Yes.

Wood: Is this like a demil process?

Mendez Romero: Yeah, we were in charge of that. And that's when Operation Southern Gambit started.

Wood: Uh-huh.

Mendez Romero: So we got back to the FOB, the company FOB, and we got tasked out to do security to the west side of the Musa Qala *wadi* (or valley).

Wood: Mmm-hmm.

Mendez Romero: Making sure no surprise attacks or anything would leak out.

Wood: So you were put out in like a platoon? As a platoon or as a squad?

Mendez Romero: Our own squad.

Wood: Were you out there for— Were you patrolling or did you set up a position?

Mendez Romero: We had vehicles. We had two four-by MRAPs [mine-restent, ambush protected] and a six-by MRAP.

Wood: Okay. Can you describe what you all were doing during that time?

Mendez Romero: We would pull security and we would stop any vehicle that seemed suspicious or, at times, we would just let one go and then stop the other one at random. And we would search their cars, search the persons, and after that we would thank them for their—thank them for helping us out and we would let them go, get on with his day.

Wood: All right. Anything happen during that time that stands out to you?

Mendez Romero: No, not there, sir.

Wood: No engagements or anything like that?

Mendez Romero: No, sir.

Wood: Okay. How long were you doing that with your squad?

Mendez Romero: We did that for two days, two or three days, sir.

Wood: Okay. From there, what did you all do?

Mendez Romero: We went back to the FOB and we stood post.

Wood: And you stood post at the FOB. How long were you doing that?

Mendez Romero: I want to say a week and a half.

Wood: Okay. And then keep walking through what did you do?

Mendez Romero: And then after that platoons would rotate out so, again, it was coming to an end. And then they set up an Afghan Police PB, and we would go there and help them out, do patrols out there. And we pushed out south, and we kept pushing out south.

Wood: Where was the Afghan Police patrol base, PB?

Mendez Romero: It was called the Ugly Hill, sir. That was just like probably two clicks [kilometers] out from Musa Qala.

Wood: Okay. And how long were you out there doing the support for the Afghan Police and patrols?

Mendez Romero: For a week to two weeks at a time.

Wood: Okay. What was that like?

Mendez Romero: We had a bond with the Afghan Army and the Afghan Police. And we did a pretty good job. They would take up quick on how to patrol with us. Their formations, they understood them. Everything was good. And when we received contact, they would help us out. They would listen; when to fire, when not to fire.

Wood: Okay. Any other particular engagements or anything happen when you were helping out with supporting the police?

Mendez Romero: Yes. We received contact in Karaway [Khiraway]. That's what it was called, Karaway. It was probably 10 clicks down south from the company FOB. And it was my team and another team from our squad going down on the west side of the wadi. We started seeing suspicious activity on the other side of the wadi. We saw spotters. We saw guys peeking in and out from the ridgeline. And we started moving. I had security and then he picked up and was moving— Excuse me, he had security and I picked up and I was moving. And that's when we started receiving contact. And we had [M]240s attached to us, so he opened up on the ridgeline, and then myself and the other team leader, we opened up with [M]203s [grenade launcher] to show a force. And contact stopped soon after that. And then our squad leader and a platoon sergeant eventually showed up. We organized a movement to go check that out, and we searched a couple of compounds. We didn't find anything and we just departed from there.

Wood: Okay. So you are supporting— Your squad, platoon were supporting [Khiraway] for how long roughly?

Mendez Romero: We were just out there patrolling, just expanding our security bubble for a week.

Wood: Mmm-hmm. All right. From there, where did you go next?

Mendez Romero: From there, Operation Western Gambit was coming up. We prepared for that. My squad was in charge of escorting the main element, or the headquarters element. And we prepared VICs [vehicles]. Other squads prepared for helo [helicopter] insertion and other squads prepared for like any kind of situation.

Wood: All right. Did you participate in Western Gambit?

Mendez Romero: Yes, I did.

Wood: Can you talk about how that went?

Mendez Romero: It went pretty good. We heard it was going to be a heavily hostile environment. So we went in prepared. [Pause] I was inside the VICs. I was in a four-by at first and then I moved over to a six-by. We had security. We provided a screen for the other element to move and flank them, flank the village. They cleared out and we remained like overwatch in a way from outside. We had a view of everything. That area was called the "Triangle of Guilt."

Wood: How long did Western Gambit go for?

Mendez Romero: It didn't go for too long. It went on for maybe two to three weeks.

Wood: And that lasted two to three weeks. And so you continued to provide your fireteam, your squad provided the security throughout Western Gambit?

Mendez Romero: Yes.

Wood: Any particular incidents, things that stand out to you from that time?

Mendez Romero: On our village that we were tasked to clear out, it wasn't— It was mostly abandoned and the village next to us that first platoon was in charge of clearing out, they were receiving heavy contact, but we didn't see any of it. We remained focused on our village in case they tried sucking us into that scenario and tried surprising us in the other. So we didn't really see any until like the fourth day; we started receiving contact from the ridge. But there were little caves on the side of the mountain so it was really hard spotting them. We didn't receive any casualties.

Wood: Okay. So that continued through Western— That was pretty much how things went with Western Gambit?

Mendez Romero: Yes. And then after we got done clearing out all the villages, the headquarters element, they were organizing there again the point of security on the PB that there was going to be there. And one day, we received a recoilless rifle shot. It hit right outside the compound that we were in. But we didn't receive any pop shots, and that was it.

Wood: Okay, nothing serious happened from what it sounds like. Okay, so you're probably about halfway through your deployment I'm guessing now, right?

Mendez Romero: About there, sir.

Wood: Can you continue to talk through some of the things that your platoon and squad did?

Mendez Romero: Then after that we came back to the FOB and [pause] then we started getting ready— Well, we were still on our post pulling security on the FOB and we got the notice that we were going to do Operation Double Check. And we started preparing for that.

Wood: All right. And what did you all do in Double Check?

Mendez Romero: In Double Check we went into Regay. And we got heloed in, sir. All platoons did. We cleared out— Our platoon cleared out the first objective, which our squad was in charge of going in and searching. And the other squads were in charge of inner and outer coordinates. And we took two elders and two like military age males with us. By that time, it was like six in the morning. The sun was just coming up. And we were walking to where our headquarters was at. And my team, we were in charge of the rear, so we started receiving contact. At first, it was just pop shots and then eventually they started getting closer. We engaged and then we turned around and headed back to headquarters.

Wood: Ever find out who was shooting at you or anything?

Mendez Romero: We never found out who it was.

Wood: So how long was Double Check for?

Mendez Romero: Double Check, from what I know, it took like maybe a month to complete. Because we put three posts at first, and then one was in Kuchnay Regay and then one was in Regay and there was another one, Solzamrabat, [Sultan Rabat] I believe. And my squad, we had our own position in Kuchnay Regay, which was a squad PB. And the platoon, the rest of the platoon was in Regai down south. And second platoon was in charge of Solzamrabat. From there we started doing— Well, before the posts were finished we stood out in the— Not in the open but on top of a hill and we did— That was our PB for the moment, for five or six, five to seven days. It was a cold, cold experience. Marines were sleeping in body bags, anything they can get a hold of to stay warm.

Wood: That was roughly what month?

Mendez Romero: That was January. And out of that little hill, our platoon called it a Winning Hill; from Winning Hill we went to— We were tasked out to clear out Objective Charlie, which was three to four clicks away. And from those objectives we have— From that objective, we have other little compounds we had to just touch; like take a look. So we took— We stepped out around, I want to say, 2000 [that night]. We didn't get back until 0500, I believe. It was a long, long night.

Wood: Yeah, I bet. Yeah. Anything else stand out to you from that time?

Mendez Romero: The locals were— They understood what we were doing, so they helped us out when we went in there and searched. We were respectful. They were respectful. And it was just a smooth process. We didn't find anything.

Wood: All right. From there, what was the next either operation or what were you doing?

Mendez Romero: From there, I said my squad was in Kuchnay Regay, right? On the Kuchnay Regay post? And we were doing patrols from there. We were in charge of making sure Route Yellow was not being IEDs emplaced or anything. So security every morning, every morning like around 0700 we would go out with the Afghan Police and just patrol down the road, making sure nothing unusual or anything. And then from there, we would do resupplies to Regay to the south post to where the platoon was at just to remain so

Defense Imagery 513838, photo by Cpl Tommy Bellegarde

Marines and Afghan police secure southern Musa Qala and work toward long-term stability during Operation Double Check.

the security bubble was strong. And then we did what was supposed to be a short patrol with the Afghan Police. That was my team and their other team that were attached to the Afghan Police. And we started taking pop shots. And at first, it seemed like nothing, well, from my perspective, it was just like pop shots like usual. In a way, you get used to it. The other team— We were on the west side of the compound; the other team was on the east side. And I believe they were taking heavy contact and my team wasn't. And I didn't have a radio at the time. So, they called us over and we started receiving— The pop shots started picking up and we decided just to start heading back and that's when one of my guys got hit in his SAPI [small arms protective insert] plate. He just got bruised and we called in air. We dropped a 500-pound bomb and we did a PSA [post support activity]. We found rifles, I believe shells, mortar shells, pistol mags. We found like hostile equipment. And then we just retreated from there, and then that was it for that patrol. Then drivers and, I believe, some fireteam leaders, they got to go back to the FOB and pick up VICs to come start the process of rotating platoons in and out, and so we were attached to headquarters at the time. We did a patrol down to Regay, and it was like a platoon minus, almost a whole platoon was there. We started receiving pretty accurate fire at first and then it just picked up. We started engaging. We couldn't really maneuver on them because they were on the ridgeline. We were on the bottom of the hill, so we just tried to do our best. We eventually pushed them out. We started RTBing [returning to base] and they got us good. They like headed back to the ridgeline and all the teams were pretty much aligned from the ridgeline so they were just shooting at one team, and from that— In case that missed, it just went down the line. So it was like— I forgot the word but it was pretty heavy contact. No matter what, you saw rounds impacting all around you.

Wood: How long did that last?

Mendez Romero: It felt like forever, but I believe like 20 minutes.

Wood: What did you all do to address that situation?

Mendez Romero: My team, we were on the farthest side from the contact, so I made sure my team made a screen from any other ambushes or anything. And the other teams started maneuvering, started trying to get closer to the contact or to the enemy. And eventually they retreated and then we just RTBd.

Wood: So that was still part of— Was that part of Double Check?

Mendez Romero: Yes.

Wood: Okay. From there, what did you do? More activity with Double Check or—

Mendez Romero: We would rotate platoons in and out, and we came back one more time. But we kept it just really local. Because it was at the end of our deployment, so we didn't want to risk it just going out.

Wood: So that was— Double Check was more or less at the end of your deployment?

Mendez Romero: Yes. It was two months maybe, or a month.

Wood: Anything else happen after Double Check that stands out to you?

Mendez Romero: Another platoon got tasked out to do another post south of Regai but I don't have too much detail for that.

Wood: Okay. Roger that. That's good. What about your work with the— You said you patrolled with the Afghan Police. Can you talk about what it was like to work with the Afghans?

Mendez Romero: At first, it was hard to communicate with them. It's like you explain something to them, and all they do is shake their head and say, "Yes, yes, yes." And then you let them do it, and they do the opposite of what you tell them. So it was a little hard, communication. And then sometimes our interpreters wouldn't translate right or how would you say? Or probably just wouldn't translate at all. We didn't know what they were saying. But other than that, on patrols, they would pick up right. They had good leaders. Some of them probably had already worked with Marines before, because we would do wedges and they would make their own wedge or just follow our trace and help us out; pick out the suspicious activity or just talk to locals and get any pointers.

Wood: Did you do anything with the Afghan Army?

Mendez Romero: Yes. That was the Afghan Army and the Afghan Police was almost— They were pretty much the same. They were good. Good help out there.

Wood: What about your interaction with the local populace?

Mendez Romero: I didn't really talk too much of elders or anything. But the kids, those would be the ones that'd come up to you the most, asking you for pens, candy, chocolate mostly, or your sunglasses. And sometimes if you're not careful, they'll just snatch it off of you and run away.

Wood: What did you think? How did they look at the Marines?

Mendez Romero: Some of them actually— We had a patrol on Regay and kids came up to us and they kind of knew English so we kind of like just words, little words we would communicate. But they said Marines were good and the Taliban was no good for them or anything. But they appreciated our help. And then the kids would help us sometimes point out the IEDs or whatever there would be.

Wood: All right. What were some of the biggest accomplishments for you and your unit during that deployment?

Mendez Romero: The biggest accomplishment, that I believe in, was all my Marines came back alive. That was the biggest accomplishment. And just the experience. We did everything successfully.

Part III
Every Marine a Rifleman, 2012

Setting Stakes by Michael D. Fay.
Marine Corps Art Collection

By January 2012, the Marine Corps' reputation was bruised and the Afghan War effort was set back when video footage emerged of Marines urinating on three Afghan corpses, one of several incidents involving Coalition troops that outraged many Afghans and Americans. In February, U.S. Defense Secretary Leon E. Panetta announced the intention to end combat missions in Afghanistan as early as mid-2013 and to assign American troops in an advisory role until their withdrawal in 2014. On 12 March, I Marine Expeditionary Force (I MEF-Forward) relieved II MEF (Forward) as RC-Southwest.

Despite the controversies and announced troop drawdowns, throughout 2012, Marine units continued to rotate into Afghanistan and continued to conduct raids and patrols throughout the Marineistan provinces, suppressing the poppy harvest and eliminating Taliban caches and sanctuaries.

In late 2012, the Taliban proved that it was still dangerous when it launched a successful raid on Camp Bastion, an airfield and logistics base northwest of Lashkar Gah, Helmand Province. They killed two Marines, destroyed six McDonnell Douglas AV-8B Harrier IIs, and badly damaged two other Harriers from Marine Attack Squadron 211 (VMA-211). The Taliban fighters were all captured or killed in the attack, during which Marine aviation personnel fought as infantry, a role Marines of this squadron last performed on Wake Island during World War II.

An India Company leatherneck pauses briefly to talk with village children while patrolling near Sangin.

Photo by Cpl Ed Galo

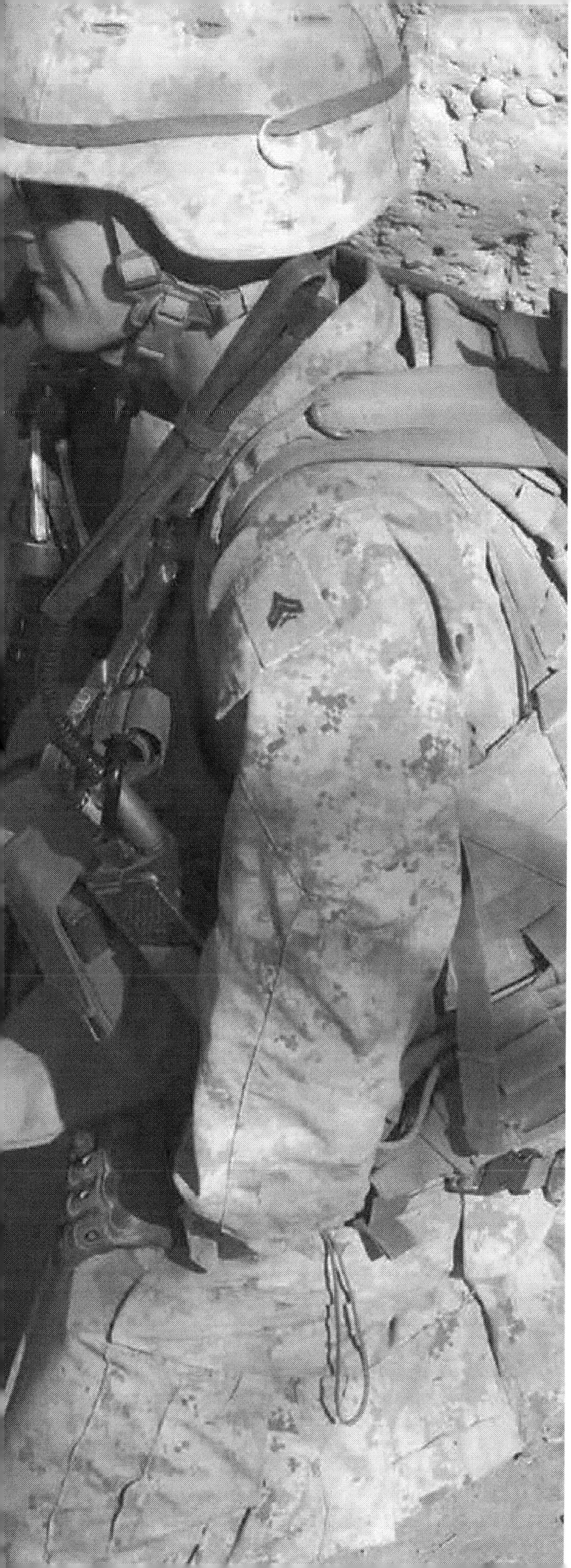

Regimental Combat Team 6: "India" Company, 3d Battalion, 7th Marines, Builds Relations in Afghanistan

by Ed Galo

Leatherneck, May 2012

The Marines of India Company, 3d Battalion, 7th Marine Regiment, maintain a great relationship with the locals in the village outside their patrol base near Sangin, Afghanistan.

They do this through daily security patrols throughout the area and by communicating with the people.

"The big thing with today's patrol [19 January 2012] was talking with the people about their farms and animals," said Corporal Jacob Marier, a squad leader from St. Louis [Missouri]. "We asked them about what they grow, what type of fertilizer they use, what their most important crops are, what their most important animals are and about how they get their water for their crops."

While Marier and part of his squad were talking to the villagers, other squad members continued to patrol throughout the village.

"We started going through the city looking for

Photos by Cpl Ed Galo

Being highly visible on security patrols around a small village near Sangin (top photo), and sitting down to discuss issues with villagers as India Company squad leader Cpl Jacob Marier is doing (bottom photo), has gained the confidence of locals.

other possible crossing points or choke points to watch out for," said Corporal Mark Yenalavitch, assistant patrol leader, from Victorville, California.

When the Marines conduct patrols, they usually are accompanied by soldiers from the Afghan National Army [ANA] as part of the turnover process from American to Afghan security forces.

"You can see the transfer from the Marines to the ANA," Marier said. "That's definitely the most rewarding part of this deployment. They are definitely showing an improvement from when we first got here. They're working harder, getting better, and taking it more seriously. This shows that they'll be able to handle things when we leave."

Note

Leatherneck 95, no. 3 (March 2012): 32–33. Reprinted with permission. Copyright *Leatherneck*.

About the Author

Corporal Ed Galo was a Marine combat correspondent deployed to Afghanistan.

Marines with Weapons Company, 1st Battalion, 1st Marines, RCT-6, patrol through Trek Nawa, Afghanistan, on 16 August 2012. Trek Nawa, an area between Nawa and Marjah Districts, contains known insurgent sites. Weapons Company has been involved in numerous firefights since the start of their deployment in June 2012.

Defense Imagery 120816-M-PC317-024

First Time, First Firefight—Marine Stays Focused during Combat

by Corporal Timothy Lenzo

Defense Video & Imagery Distribution System, 21 August 2012

Patrol Base Detroit, Afghanistan—It is a moment of truth for many Marines—the first time they are in combat and their training is put to the test.

When his squad took enemy contact during a recent patrol through Trek Nawa [Helmand Province], Private First Class Timothy Workman found his moment. "I could hear rounds cracking over my head," said Workman, mortarman [in] Weapons Company, 1st Battalion, 1st Marines, Regimental Combat Team 6. "The adrenaline started pumping right away."

This was Workman's first firefight. A year ago, he was standing on the yellow footprints at Marine Corps Recruit Depot Parris Island [South Carolina] to start basic training, a tradition every Marine recruit goes through. Now, with the midmorning sun beating down, he was crouched in a ditch, his M16 assault rifle ready.

Workman and his fellow Marines fought an estimated six to eight enemies. The insurgents fired from several different positions, shooting through small holes in walls several hundred meters away. Marines took cover in mud compounds and behind mounds of debris.

Defense Imagery 656220, photo by Cpl Ed Galo

Marines with Scout Sniper Platoon and their attachments stood in an over watch position to provide surveillance and gather intelligence on the enemy during Operation Helmand Viper.

Workman's squad split into two units during the firefight. Workman, from Peebles, Ohio, went with Staff Sergeant David Simons, his platoon sergeant, as they moved forward to return fire.

"At one point, I witnessed Workman moving up into position to engage the enemy," recalled Simons, from Sidney, Montana. "He fired on the enemy, and when the enemy returned fire, it allowed us to open up with our machine gun."

The morning sounds of birds and farmers were replaced with the sudden burst of rifles and the "rat-ta-tat-tat" of machine-gun fire. An hour later, the fight was over and the Marines returned to their patrol base.

"Since we've been out here, it's pretty common for [the other Marines] to engage in firefights," Workman explained. "These [insurgents] will stick around and [fight] for awhile."

Workman's company patrols the volatile Trek Nawa area of Afghanistan. Trek Nawa is an area between the Marjah and Nawa Districts of Afghanistan in Helmand Province.

The Marines engaged enemies in firefights ranging from isolated pot shots to three-day-long battles. For Workman, the fighting hit home six months before he left for boot camp. In December 2012, his friend's older brother, Luke, was killed while serving near this same area of Afghanistan.

"I had gone to school with Luke's brother since the sixth grade," said Workman. "At the time [of

Luke's death] I had already decided to join, but this motivated me to continue the work that Luke gave his life for."

In addition to Luke, Workman said he's lost a couple other friends to the war in Afghanistan.

Despite losing friends to combat, Workman remembered his training and focused on his job during the fight.

"I was trying to get positive identification on the enemy, trying to find where they were firing from, looking for spotters and just covering my brothers," said Workman.

Now that his first firefight is over, Workman's platoon will monitor his behavior.

"The main thing we look for is a Marine's mindset after their first time in combat," Simons said. "We are a family, so we can tell when one of our brother's [*sic*] is acting differently. We watch for it and take care of each other."

Simons said Workman seemed mentally strong before and after the firefight. He is proud of Workman and the discipline he showed on the battlefield.

The Marines have more operations planned through Trek Nawa and expect more firefights before they return to the states.

"I can say there's nothing else like the Marine Corps," Workman said. "I'll continue picking my sergeants' and seniors' brains, trying to be better prepared for the next fight."

Note

This work was originally published by Defense Video & Imagery Distribution System, which is free of copyright restrictions.

LCpl Harrison York, an AV-8B Harrier II Plus maintainer with VMA-211, 3d MAW (Fwd), salutes Capt Stephen White, a Harrier pilot, as he taxis in to the Camp Bastion flightline on 26 September 2012. The aircraft bears the name of LtCol Christopher Raible, the squadron's commanding officer who was killed during an attack on the Camp Bastion airfield on 14 September.

Defense Imagery 120926-M-ZM862-002

Otis Raible Defines Leadership

by Lieutenant Colonel Michael Jernigan

Marine Corps Gazette, July 2013

First place in the 2013 General Robert E. Hogaboom Leadership Writing Contest

A Hero for All Marines

> Trained men who will stand and fight are never obsolete. It was not the bowman, but the long bow, not the cavalryman, but the horse, which vanished from the scene. Men—the man, the individual who is the Marine Corps symbol and stock in trade—constitute the one element which never changes.
>
> ~Colonel Robert D. Heinl, 1962[1]

I was sitting in my air-conditioned office cubicle stateside when I heard the news: a Marine squadron commander had been killed in Afghanistan. I read the wire and realized that I was acquainted with him and familiar with the battlefield geometry. More surprising, but not to Marines, was that he was an aviator killed in the ground defense of an attack on Camp Bastion in Helmand Province.

Marine colonel, professor of leadership, and recipient of the Navy Cross, the late John Ripley said, "There comes a time in every man's life when he is called to use his unique and special talents."[2] Saturday, 15 September 2012, was one of those times for Lieutenant Colonel Christopher

K. "Otis" Raible, when Camp Bastion's airfield was attacked. Two U.S. Marines were killed, nine others wounded; six AV-8B Harrier attack aircraft were destroyed, and two others badly damaged. Otis was one of the Marines killed.

Otis was serving as the commander of Marine Attack Squadron 211. The squadron was nearing the end of its seven-month deployment. Shortly after 2200, he was returning from chow after a day of flying and heard explosions along the flight line. Fifteen Taliban wearing U.S. uniforms and armed with rocket propelled grenades (RPGs), machine guns, and suicide vests breached Camp Bastion's perimeter and began destroying Harriers.[3] Otis immediately checked on his Marines in billeting and then, armed only with his pistol, hurried to the point of attack. He found that the attackers had split into three groups: two focusing on destroying aircraft and the third headed to kill Marines while they slept.

Official U.S. Marine Corps photo

U.S. Marine Corps aviator LtCol Christopher K. Raible.

Otis ran through enemy fire across an open area 100 yards across and grabbed a handful of mechanics, avionics technicians, landing support specialists, and bulk fuelers—the usual crew one would expect to find on an operating flight line, but not the usual crew one would likely find in a quick reaction force. However, leadership is mostly the same whether applied to airmen or infantrymen, and Otis organized a hasty counterattack. He split his small force into two groups. One engaged the Taliban headed toward main side Camp Bastion while Otis led eight Marines to engage the enemy in a maintenance hangar.

Otis' aggressiveness forced the Taliban to abandon their attack plan and seek cover, which in turn allowed dozens of Marines to escape an indefensible position, setting a more secure firing position and limiting the enemy's advance. [Bell] AH-1W Cobras and [Bell] UH-1Y [Iroquois] Hueys also stationed at Camp Bastion began to take off amidst the flames of burning aircraft. While Otis and his team engaged the insurgents in the hanger [*sic*], the airborne helicopters were able to use their night-vision equipment to spot the enemy and begin "danger close" gun and rocket runs.* Soon, more Marines arrived at the point of necessity and the defense began to have backbone. One dismounted helicopter crew fired a [M]240G [medium] machine gun at RPG flashes, while other Marines engaged the Taliban from a bus normally used to transport flight crews to chow. The fight raged fiercely for four hours until 14 of the insurgents were killed and the remaining 1 captured.

* According to the Department of Defense's *Dictionary of Military and Associated Terms*, the term *danger close* indicates that friendly forces are within close proximity of the target and the incoming close air support, artillery, mortar, and naval gunfire support fires.

The Marines and their British counterparts spent the rest of the night ensuring that there were no more infiltrators, but the battle was over. Sadly, at some point during the firefight, Lieutenant Colonel Raible and Sergeant Bradley W. Atwell were both killed by RPG shrapnel.

The attack was significant, with the *Wall Street Journal* describing it as "the biggest single-day loss of U.S. combat aircraft since the Vietnam War."[4] Major General Charles M. Gurganus, commander of Regional Command (Southwest), characterized it as well planned, well rehearsed, and "not conducted by a bunch of yahoos."[5] Reports indicate that Otis Raible's heroic actions and instinctive rallying of those few Marines at hand were the key factors in establishing an effective defense. His quick thinking and aggressive leadership kept the enemy from infiltrating throughout the base and prevented the night massacre the Taliban had planned.

Otis' squadron responded to his death with resilience. They fought on after his death, repelling the attack. They used replacement aircraft and completed their deployment. One Marine injured in the attack told a reporter, "My commanding officer never feared death and would want us to keep fighting. That's what he would do."[6] A sister squadron commander said simply, "Otis trained them well."[7] Those two comments are marks of a great leader and form perhaps the highest praise a commander could receive from his subordinates and peers; together they say, "We did well because he trained us and led us."

Otis' example illustrates several traits of other known leaders. Civil rights leader Martin Luther King Jr. believed that the ultimate measure of a man is not where he stands in moments of comfort, but where he stands at times of challenge.[8] At the key point in the battle, Otis stood where all commanders should wish to be—in front of their Marines, leading them in repelling a determined enemy. By all accounts, his personal leadership was one of the critical factors in reducing the loss of human life.

Gene Krantz was a flight director for NASA [National Aeronautics and Space Administration] in 1970 and [was] considered largely responsible for the safe return of Apollo 13 after an oxygen tank exploded in midflight. In his estimation, "The leader has to be unflappable. No matter what is going on around you, you have to be cooler than cool. You have to be smarter than smart."[9] Krantz could have been talking about Otis Raible at [Camp] Bastion. He was calm when others panicked; he was organized while chaos swirled around him; he acted, and his action inspired others to act. That description provides as good a definition of leadership as one can find.

The conduct of the battle certainly adds to the legacy of the Marine Corps, validates the training Marine officers get at The Basic School, and reinforces the axiom "every Marine a rifleman." But equally importantly, Otis Raible's last night serves as an example to all combat leaders. He did not happen into it—he prepared for it.

Otis, like many commanders, published to his subordinates what was important to him. His "Commander's Guidance for Squadron Attack Pilots" contained typical themes appropriate to aviators: "If flying jets and supporting Marines is your passion and your profession, you are in the right squadron. If you average one hour per work day studying, six months from now you will be brilliant. When all else fades away, attack pilots have one mission: Provide offensive air support for Marines."[10] That last one is interesting because, in addition to "providing offensive air support for Marines," Otis clearly was also effective at conducting ground defensive operations with Marines.

I believe other tenets of his core leadership

Defense Imagery 774229, photo by LCpl Uriel Avendano

LtCol Raible's VMA-211 is based in Yuma, AZ.

beliefs were in action on 15 September. In his guidance, Otis reminded his team, "Keeping things 'simple and easy to execute' will usually be your surest path to success." Otis knew that simplicity is an antidote for the chaos of combat. His defensive plan that night was the simplest of all: "Run to the sound of the fight, rally the Marines you find there, and engage the enemy." This is certainly not the technique to win a fictional tactical decision game, but exactly the right application for the realities of a firefight.

Otis wrote that it was important to "hire for attitude and train for skill." Clearly, both attitude and skill were qualities in abundance during the defense of Bastion. Otis demonstrated these traits personally, and his squadron emulated him, displaying them collectively.

Toward the end of his "Commander's Guidance," Otis directed his pilots to ask themselves this question: How good could I be if I really gave this my all? To this I would answer, "Otis, you were really good—good enough to save your Marines and good enough to serve as an example for the rest of us." The Marine Corps itself is better because of the brief time Lieutenant Colonel Chris "Otis" Raible was in it. Chief Warrant Officer 4 Hershel W. "Woody" Williams, the last living Iwo Jima recipient of the Medal of Honor, expresses it well: "You can be a hero on one day, but to be a role model is a lifetime thing."[11]

Otis, you are my hero, and a role model to many.

Note

Marine Corps Gazette 97, no. 7 (July 2013): 13–15. Reprinted with permission. Copyright *Marine Corps Gazette.*

About the Author

Lieutenant Colonel Michael Jernigan commanded a battalion in Afghanistan and served as the senior American at Camp Bastion from 2008 to 2009.

Endnotes

1. Robert D. Heinl, *Soldiers of the Sea: The United States Marine Corps, 1775–1962* (Annapolis: Naval Institute Press, 1962), 542.
2. Gen James L. Jones, 32d Commandant of the Marine Corps, "Birthday Message" (memorandum, Headquarters Marine Corps, 2001).
3. Gretel C. Kovach, "Marines Recall 'Surreal' Attack at Afghan Camp," *San Diego Union Tribune*, 6 October 2012. Much of the account of how the battle developed comes from this article.
4. Nathan Hodge, "Coordinated Taliban Raid Penetrates Base," *Wall Street Journal*, 16 September 2012.
5. Jim Michaels, "Taliban Trained Inside Pakistan," *USA Today*, 28 November 2012, 1.
6. Kovach, "Marines Recall 'Surreal' Attack."
7. Ibid.
8. Joel Comiskey, *Leadership Explosion: Multiplying Cell Group Leaders to Reap the Harvest* (Houston, TX: Touch Publications, 2000), 47.
9. Michael Useem, *The Leadership Moment: Nine True Stories of Triumph and Disaster and Their Lessons for Us All* (New York: Random House, 1999), 84.
10. Christopher Raible, "Commander's Guidance for Squadron Attack Pilots" (memorandum, Yuma, AZ, 9 December 2011).
11. Jones, "Birthday Message."

Part IV
Advising and Nation Building, 2013

Returning Sniper Fire by Michael D. Fay.
Marine Corps Art Collection

Marine General John R. Allen, commander of the International Security Assistance Force (ISAF) and U.S. Forces Afghanistan and senior American commander in Afghanistan, was replaced on 10 February 2013 by Marine General Joseph F. Dunford Jr.

On 28 February, II Marine Expeditionary Force (Forward) relieved I Marine Expeditionary Force (Forward) as Regional Command-Southwest (RC-Southwest). Marines continued counterinsurgency and training operations throughout 2014, turning over responsibility for security operations district by district to Afghani forces.

On 18 June, NATO handed over operations to the Afghan military. Marines, along with the rest of the American forces still in Afghanistan, focused on training missions.

1stLt Brent Bonnema with the Afghan National Civil Order Police, Kandak 1 Advisor Team, RCT-7, demonstrates firing the pistol while conducting small-arms, live-fire training on 25 April. The Marine advisor team is supported by RCT-7's ANSF liaison cell.

Defense Imagery 929123, photo by Cpl Alejandro Pena

In Time of Transition, Marines Support Afghan Advisors

by Sergeant Ned Johnson

Leatherneck, July 2013

Helmand Province, Afghanistan—The mission in Helmand Province, Afghanistan, has forced a change from conventional Marine units to smaller units who advise Afghan National Security Forces (ANSF). These advisors are formed into Security Forces Assistance Advisor Teams (SFAAT), [which are] much smaller than a Marine company. Without the standard setup of a battalion, these Marine teams require a different type of support.

Marines serving with the ANSF liaison cell with Regimental Combat Team 7 [RCT-7] provide support to the more than 20 teams within the RCT's area of operations. "The Marines in the cell are responsible for the initial training each SFAAT receives when arriving in Afghanistan, as well as follow-on training throughout the team's deployment," said Gunnery Sergeant Matthew Lockwood, the cell chief.

The training includes live-fire exercises involving movement-to-contact drills and shooting under stress. The advising mission creates a need for more advanced training because the 20- to 30-man teams often operate independently.

"After an assessment of the training package,

Defense Imagery 948655, photo by Cpl Kowshon Ye

U.S. Marines with Fox Company, 2d Battalion, 8th Marines, RCT-7, patrol through a village during Operation Nightmare in Nowzad, Afghanistan.

Marines added an insider threat training program to help prevent green-on-blue attacks against U.S. advisors," said Lockwood.* The insider threat training teaches the team members how to better recognize potential threats and dangerous scenarios.

To establish relationships and build rapport, the Marines also visit the SFAATs to ensure they are well-equipped and to gain a better understanding of the individual Afghan force being advised.

"Unless you're on the ground, you don't really know much about what is actually going on," said First Lieutenant Jeremy Prout, a liaison officer with the ANSF cell. "We get out there and see what's working and what's not."

* *Green-on-blue attacks* refer to attacks on Coalition troops by Afghan forces.

The Marines operate as administrative and logistic liaisons between the teams and the RCT, providing team-specific gear, equipment, and even food. At one point, the cell sent more than 50 care packages to the Marines to provide them with snacks and other amenities not available at the smaller bases.

"Along with being liaisons, the Marines in the cell also are the ANSF subject-matter experts for the RCT," said Prout. This includes briefing the RCT commanding officer on the effectiveness of Afghan forces in each district.

While the idea of advising Afghan forces is not new, the liaison cell is. Regimental Combat Team 6, which was replaced by RCT-7, operated a one-man liaison cell, while RCT-7 operates a six-man cell.

"The one Marine did a great job, but we have been able to provide more support [through the work of more Marines]," Prout said.

As the advising mission continues, the ANSF liaison cell Marines will continue to assist the teams.

"We've been able to create a great connection with the teams," Lockwood said. "Whenever we have visited them, they are always very thankful for what we've been able to do together."

Note

Leatherneck 96, no. 7 (July 2013): 16–17. Reprinted with permission.

About the Author

Sergeant Ned Johnson was assigned as a combat correspondent with II Marine Expeditionary Force (Forward).

While on patrol in Helmand Province in late December 2013, LCpl Patrick Tomassi, a grenadier with 1st Battalion, 9th Marines, sees firsthand the challenges of fighting an insurgency.

Defense Imagery 1146361, photo by Cpl Austin Long

This Is My Life: Marines in Afghanistan

by Corporals Austin Long and Paul Peterson

Leatherneck, March 2014

The war in Afghanistan is the longest sustained conflict in American history. For more than 12 years, Marines have cycled in and out of the country. Most Marines today have never known a time when deployments did not loom on the horizon. It has become a facet of their lifestyles, and it has shaped the people who lived through it.

Now, as the war in Afghanistan comes to an end, four Marines with 1st Battalion, 9th Marine Regiment, operating in Helmand Province, share a little bit about who they are as members of that select community. Their attitudes, leadership styles, experiences, and reasons for serving are different. They are honest and hopeful, rancorous and rash, proud and blunt. Three are combat veterans. One is serving his first tour overseas.

A squad leader, team leader, grenadier, and scout leader—they have all weathered enemy fire during this deployment. For the most part they are where they want to be, somewhere between loving their jobs and simply enduring the miseries of deployment.

Sergeant Bryan W. Early
Squad Leader

At 25 years old, this is Sergeant Early's third deployment. He uses his past experiences to lead

Defense Imagery 1146392, photo by Cpl Austin Long

Sgt Bryan Early, a squad leader with 1st Battalion, 9th Marines, RC-Southwest, directs a Marine where to provide security while on patrol in Helmand Province on 19 December 2013.

his Marines today. A Libby, Montana native, the Marines know Early by his natural, happy mood. He tries to think of his men as little brothers to remind him he is responsible for the lives of other people's sons. He has a wife and a 2-year-old son.

Q: What thoughts run through your head when planning patrols?

A: The other squad leaders and I sit down and focus mostly on the safety of our Marines. When we make our plans, we look at all the intelligence we have and plan around that while also keeping the commander's intent in mind.

Q: When you're going through a firefight, what kind of squad leader are you?

A: I try to be as aggressive as possible. I know that's hard for some people to wrap their heads around, but as an infantryman our mission is to locate, close with, and destroy the enemy with fire and maneuver. So I try to instill that as much as possible in my Marines. If I show them that I'm scared, then they're not going to be willing to follow me into that gunfire, but if I'm aggressive and happy and I'm the man to step out into the fire, then that shows the Marines that [leading from the front] concept and makes them more willing to follow me.

Q: Do you approach everything with that aggressive mentality?

A: Yeah, especially in the Marine Corps. I try to be as aggressive as possible. But life outside of the Marine Corps, you've got to step back and not be as aggressive. I've tried the aggressive approach, and it normally doesn't go over well. [Laughing]

Q: Why are you the first one through the door and in front of all the other Marines?

A: I'd rather be the first one through the door, so that if anyone has to take a round, I'd rather it be me. You put so much hard work in training these guys; you want to give them the opportunities to not only succeed in the Marine Corps, but in life. If anyone has to take the bullet, I'd rather it be me. This is my third deployment, and I have six plus years in the Marine Corps. I've already had my time.

Q: How does it feel watching your Marines operate on patrol?

A: I saw it when we took our first contact [with the enemy] that it was muscle memory for these guys. They never hesitated or choked, and that's the best feeling to have.

Q: Do you regret coming into the Marine Corps right after high school?

A: Absolutely not.

Q: Why is that?

A: Being a Marine was one of my dreams as a small child. My mom has pictures of me when I was four or five [years old] walking around in old Marine Corps tricolors [utilities] with a pellet gun at shoulder arms.* [Laughing]

* According to the manual of arms, shoulder arms (right or left) is a standing or marching position whereby the weapon rests sights up on the shoulder with a hand on the butt of the weapon.

Q: What kind of camaraderie do you have with your junior Marines?

A: One of the big things I learned as a junior Marine is that you train your Marines like they're your little brothers. Always keep in the back of your mind that's someone's son. Treat your Marines accordingly, and they'll give you the respect that you deserve. If I get hit, I don't want them running out there to get me just because I'm another Marine. I want them running out there to save me because I'm Bryan Early.

Corporal Charles A. Kristel
Team Leader

He is blunt and authoritative but also respected. Corporal Kristel, a Schenectady, New York native was wounded during his previous deployment to Afghanistan. A stocky, deep-voiced man of few words, many unrepeatable in print. Kristel joined the military at the age of 20. After only three-and-a-half years in the Marine Corps, he has elevated himself to a position of leadership through strength of will, demonstrated competence, and unbending character.

Q: What were some of the traits you saw in your leaders coming up in the Marine Corps that made them successful?

A: I would say doing whatever's necessary. They made a point to do their job in the best manner possible and just succeed, as opposed to being mediocre because mediocre usually fails.

Q: You hold a pretty senior position. How did you get to that point?

A: Being awesome at my job.

Defense Imagery 1146379, photo by Cpl Paul Peterson

Cpl Charles Kristel, a team leader with 1st Battalion, 9th Marines, uses his radio to coordinate with another section of Marines during an interdiction and clearing operation in Helmand Province on 19 December 2013.

Q: Did you ever picture you'd be who you are now?

A: Yeah, this is pretty much who I was before I joined.

Q: What were conditions like during your first deployment?

A: It was a good time. It was considerably more miserable than this. Week one, we [moved into] a compound and lived in there for four months in north Helmand. It rained a lot and was very uncomfortable.

Q: Why do you like working with the infantry?

A: Well, the infantry is the backbone of the Marine Corps. It's what defines it.

Q: There's a lot of trust placed in you and other noncommissioned officers on patrol. How do you earn that trust?

A: Through my conduct—it should be pretty evident that I know what's going on. I make a point to make it obvious I'm not an idiot.

Q: Would you say that you're a confident person?

A: Confidence lets me know I'm making the right choice.

Q: How do you delegate responsibility?

A: You delegate accordingly. [New Marines] only listen to you if they have some modicum of respect for you. So you need to distinguish yourself as either someone who knows what's going on or somebody who doesn't.

Q: How do you share leadership and break up responsibility?

A: We just converse. It's a matter of what needs to be done and who can get it done the best.

Q: How often do you lead people?

A: Every day.

Q: How do you encourage leadership in others?

A: It's always said, make them into leaders. People who obviously distinguish themselves as being competent, they will naturally grow into leaders. It's not about the people above them.

Q: If you recognize you've got somebody who's a strong leader, what do you do?

A: Let them do their thing.

Q: What's more important, your rank or your personal authority?

A: Your authority. I'm filling a sergeant's billet.

Q: How did you become versed in all those things that just seem second nature today?

A: Well, when I was [new], I just made an effort to learn as much as I could so I would be efficient as I picked up rank.

Q: Is there satisfaction for you in leading?

A: When things go well.

Lance Corporal Patrick Tomassi
Grenadier

An Odessa, New York native, Lance Corporal Tomassi cannot seem to stop smiling, even when told to. He has been in the Marinc Corps for two years, and this is his first deployment to Afghanistan. Being the new guy, he is often called on to complete miscellaneous projects. His goofy smile fades from time to time, but Tomassi's go-to-it attitude and optimism carry him.

Q: What do you think about being so young in the infantry on your first deployment?

A: I'm 20 years old, I'm in the Marine Corps, and my job is very important. I have a bunch of responsibilities on my shoulders. I've always been told just because you're younger you can't do this. But out here, we're proving them all wrong. I've got 17- and 18-year-old friends here holding [rifles], running toward gunfire. I'd rather have those guys protecting my back than anybody else.

Q: When you're going on a patrol, what are some of the thoughts going through your mind?

A: The safety of the guys around me and doing my best to assure we return with the same guys uninjured.

Q: What kind of connection do you have with the guys you work with?

Defense Imagery 1146338, photo by Cpl Austin Long

LCpl Patrick Tomassi remains on alert while on patrol in Helmand Province in late December 2013.

A: I definitely have a strong connection with all the guys. I've known these guys since January [2013] when I came to 1st Battalion, 9th Marines. We've been through [extensive] training and all the ups and downs of the Marine Corps.

Q: Do you self-criticize or take critique from others and apply that to the next patrol?

A: I try to learn from my mistakes and the mistakes of others just so nothing bad happens the next time, and I try to use that to better myself and the other Marines I'm with.

Q: What have been your favorite parts on this deployment?

A: Hanging out with the guys and all the experiences we've gone through. Even if it's a bad time, it's an experience I'm able to share with the guys. I wouldn't have it any other way. These guys are my family. I couldn't see myself doing anything without them.

Q: What thoughts run through your head when you guys get fired at by the enemy?

A: The first thing is get down [smiling] and then try to find out where the fire is coming from. Hopefully, we get through this, take down the enemy, and get home safe.

Q: What type of camaraderie do you have with the Marines?

A: Going through the worst times possible has

brought us close. We were [training in] Bridgeport, California, and it was from -15 to 15 degrees, and we went through that together.* Having to rely on each other to survive brought us close.

Q: Would you take a bullet for these guys?

A: I'd definitely take a bullet for these guys, in a heartbeat.

Q: Why?

A: They come before me. Most of these guys have a wife and kids. And that's why I signed up, to protect them.

Q: Could you imagine doing anything else?

A: I couldn't imagine doing anything else. I'm protecting and serving my country. I'm doing what I like to do, and just knowing everyone's back home safe and that I'm here just makes me feel a lot better.

Corporal Dennis Cox
Scout Sniper

This is Corporal Cox's third deployment and his second to Afghanistan. The New York City native enlisted at the age of 18, inspired in part by the events of 9/11. He left for his first deployment as a rifleman. Now, at the age of 24, he has returned to Afghanistan as a sniper, supporting Marines conducting foot patrols in Helmand Province. Cox is both serious and relaxed but fun loving at the same time. He wants to continue his line of work after the military.

Q: Why did you come into the Marine Corps?

A: I've always wanted to do it. I always wanted to do something as far as serving my country.

Q: Did you ever think of doing anything else?

A: Not really.

Q: What's unique about your job that sets you apart from other people?

A: I have the long gun, and I can see a lot farther. I remember when I did my first deployment to Afghanistan I was a [rifleman]. I was like, "I can barely see these people. It was like shooting at little dots."

Q: What about your job do you like most?

A: Operating. I love being out there. I don't know. It sucks that your boys are out there with you, but it's like brotherhood through misery. It definitely helps. You don't want to go through that experience alone. Not only out here, but even in garrison, you have a lot of stuff you have to deal with. You just deal with it together.

Q: Have you seen yourself grow over the last several years?

A: Oh yeah, I've definitely got[ten] thicker skin. I never really changed, just got thicker skin.

Q: Have you seen how you fit into the bigger picture?

A: [We] definitely instill some fear in the enemy. When you can get that close shot to them, and they watch [their guy] go down, it's like, "We got

---

* Bridgeport, CA, is the site of the Marine Corps' Mountain Warfare Training Center.

Defense Imagery 1146368, photo by Cpl Paul Peterson

Cpl Dennis Cox, a scout sniper with 1st Battalion, 9th Marines, on his third combat deployment, uses his rifle scope to scan distant buildings during an interdiction operation in Helmand Province on 19 December.

to back off because these guys can reach out and touch me."

Q: Could you imagine doing anything else other than what you do now?

A: Not really.

Q: Have you found the Marine Corps [to be] a place you really fit into?

A: I definitely think I was like this before the Marine Corps. I've kind of grown, as far as dealing with things. You can't just complain about everything. You just have to truck on and roll with the punches. It kind of helps you in life too.

Q: Have you had people who inspired you along the way?

A: I've had [them] all throughout my Marine Corps career. You always take the good things from people and just try and instill them in yourself. If you see the bad things, you're just like, "Hey, I'm never going to be like that."

Q: What are some of the traits you've seen in successful leaders?

A: [They are] kind of like one of the guys, but [they are] firm too. You don't want to upset them because you feel like you let them down, and that hurts you more than anything.

Q: What characteristics make you good at what you do?

A: I don't know. I'm just me. I'm competent. [Laughter]

Q: What's it like to go out into the field and perform your mission?

A: I love operating. Not all [missions] are home runs, and you do have some snoozers. It happens. But whenever things do go down, I love it. You get that adrenaline pumping.

Q: How do you cope with the unpleasant things?

A: We just have to vent to each other. We'll just talk to each other. That's it because our peers are all we've got out here.

Q: What kinds of people succeed in the Marine Corps?

A: Like I said, you don't want to do anything to upset them if you're under them. It's hard to explain. They have to be guys who stand their ground.

Q: How does the training and experience help you distinguish leaders?

A: We're going to find out if you're worthy or not. It's kind of like the Spartans. You need that experience because we're going to find out. Either you're going to go into baby mode or you're going to become a man.

Q: Have you found fulfillment in your job?

A: This is what I always wanted to do. It's definitely like a life-achievement goal. It's one of those things where I don't want to have to think, "I wish I did this," or something. I can grow old and happy knowing I made it.

Authors' note: the Marines selected for these interviews were chosen for their varied experience and leadership roles within their unit. None were in the military when the war in Afghanistan began, but they will be some of the last Marines to see combat in Afghanistan.

Note

Leatherneck 97, no. 3 (March 2014): 36–40. Reprinted with permission. Copyright *Leatherneck*.

About the Authors

Corporals Austin Long and Paul Peterson were assigned as Marine Corps combat correspondents with RC-Southwest, Helmand Province.

Col Austin E. Renforth, commanding officer, RCT-7, interacts with Cpl Dunn, an improvised explosive detection dog, while visiting with Marines on FOB Musa Qala, Helmand Province, Afghanistan.

Defense Imagery 817116, photo by Cpl Alejandro Pena

Last Marine Infantry Regiment Departs Helmand Province

by Sergeant Ned Johnson

Leatherneck, October 2013

Camp Leatherneck, Afghanistan—With temperatures reaching 110 degrees or more, Marines and sailors with Regimental Combat Team 7 [RCT-7] closed the final page on a chapter in Helmand Province during a casing of the colors ceremony at Camp Leatherneck, Afghanistan, 31 July.*

Over the past several years, regimental combat teams turned over their responsibilities to the next RCT as the combat mission in southwest Afghanistan continued, but this time was different. With Afghan National Security Forces having taken control of combat operations throughout Helmand, which has eliminated the need for a Marine infantry regimental headquarters, RCT-7 officially turned over its responsibilities to Regional Command Southwest.

Marines with RCT-7 spent the last 10 months providing support to ground combat units and security force assistance advisor teams and retrograding equipment out of Afghanistan.

While the RCT will be the last in Helmand Province, the Marines recognize they are the

* Casing the colors is a traditional flag ceremony that symbolizes the movement of a unit, typically either at the beginning or end of a deployment.

Defense Imagery 985194, photo by Sgt Lia Gamero

Col Austin E. Renforth (left), commanding officer of RCT-7, and SgtMaj Scott Samuels, RCT-7 sergeant major, furl the RCT-7 flag during a casing of the colors ceremony at Camp Leatherneck on 31 July. "We've thought the whole time about the legacy that came before us and not wanting to let down those Marines," Renforth said. "We're proud to have been the last regiment."

benefactors of the hard work of previous Marines.

"It's the progress that's been made that allows us to be the last [RCT]. This is a big day for all the regiments who came before us, and if it wasn't for their hard work, we wouldn't be in this position—to be able to leave," said Colonel Austin E. "Sparky" Renforth, the RCT-7 commander.

In the fall of 2009, RCT-7 began its first deployment to Helmand Province. Since then, Marines with 1st, 2d, 5th, 6th, and 8th Marine Regiments have deployed in support of Operation Enduring Freedom. In October 2012, 7th Marine Regiment again deployed and is the only RCT to have deployed there twice.

"I think the [Marines'] endurance and flexibility has impressed me the most," Renforth said. "What the Marines have done is kept their positive attitude the entire time—nothing ever fazed them, and it was really impressive to be a part of it.

"I say all the time, 'It's amazing what you can do when you don't care who gets the credit.' That's what our Marines have lived by, and they all know that they made a difference," he added.

Renforth and Sergeant Major Scott Samuels, the RCT-7 sergeant major, furled the regiment's colors and then cased them, symbolizing that the regiment's mission was complete. "We've thought the whole time about the legacy that came before us and not wanting to let down those Marines," Renforth said. "We're proud to have been the last regiment."

Note

Leatherneck 96 No. 10 (October 2013): 34–35.

About the Author

Sergeant Ned Johnson was assigned as a combat correspondent with RCT-7, 1st Marine Division.

Part V
Exiting Afghanistan, 2014

Heavy by Michael D. Fay.
Marine Corps Art Collection

Marines continued counterinsurgency and training operations throughout 2014. In February, II Marine Expeditionary Force (Forward) turned RC-Southwest over to Marine Expeditionary Brigade-Afghanistan (MEB-A).

In April, Afghanistan held presidential elections, followed by a run-off election in June. Ashraf Ghani replaced Hamid Karzai as president of Afghanistan in an historic democratic transfer of power. In May 2014, President Barack H. Obama declared that U.S. combat operations in Afghanistan would end in December of the same year.

On 26 August 2014, U.S. Army General John F. Campbell succeeded General Joseph F. Dunford Jr. as commander of the International Security Assistance Force (ISAF) and U.S. Forces Afghanistan.

In October 2014, Marines handed Camp Leatherneck in Helmand Province over to Afghan forces, and in December, NATO and the United States "ended" combat operations in Afghanistan.

Leathernecks with 1st Battalion, 7th Marines, conduct a security patrol during a mission in Helmand Province on 5 June. During the previous mission on 29 May, the company discovered a drug lab and removed more than one metric ton of narcotics.

Defense Imagery 1402194, photo by Cpl Joseph Scanlan

Marines Confiscate Narcotics, Prevent Taliban from Reaping Profits

by Corporal Joseph Scanlan

Leatherneck, August 2014

Camp Bastion, Afghanistan—Nearly a month into Afghanistan's summer fighting season, leathernecks with 1st Battalion, 7th Marine Regiment, continued a series of disruption operations in Now Zad District, Helmand Province, Afghanistan.

One of their most successful missions yet took place during a 29 May patrol in the district. Marines with Bravo Company discovered a drug lab that contained more than one metric ton of narcotics. Although counternarcotics is not the Marines' primary mission, they removed the discovery from the battlefield and prevented Taliban fighters from using profits from the sale of narcotics to purchase weapons, ammunition, and improvised explosive device (IED) materials.

Under the cover of night, the company inserted into Now Zad via [Sikorsky] CH-53E Super Stallion helicopters.

"I noticed a couple [of] spotters as we were moving into the area," said Corporal Cody Evans, a squad leader.

Spotters often are used to monitor Coalition force movements and can aid with targeting.

"About an hour after we set up our security positions, an insurgent fired a rocket-propelled

Defense Imagery 1402191, photo by Cpl Joseph Scanlan

CH-53E Super Stallions with Marine Heavy Helicopter Squadron 466 transporting Marines with 1st Battalion, 7th Marines, during a mission in Helmand Province to disrupt operations in a known Taliban bed-down location.

grenade and it exploded next to my truck," said First Sergeant Michael Grassl.

The infantrymen began maneuvering into the town, and a squad quickly discovered an emplaced IED and other IED materials. The leathernecks cordoned off the area and within minutes came under enemy fire. They tactically maneuvered toward the enemy fighters in an attempt to return fire, but the fighters retreated before the Marines could close the distance.

"We soon discovered that the enemy fighters were maneuvering through underground wells to run away after shooting at us," Corporal Evans said.

The squad continued to move north through the town and ultimately discovered the narcotics lab. The leathernecks cordoned off the area and came under enemy fire.

"The enemy definitely valued the narcotics lab we discovered," Grassl said. "They continuously tried to keep us away from the compound by engaging us with small-arms fire."

Ultimately, the Marines deterred further enemy fire and removed the narcotics, concluding the operation.

"Our success is attributed to the Marines and their hard work and efforts," Grassl said. "What we accomplished shows that steady tactical patience and steady operations pay off. It's not every day that we hit a home run, but it just happened to be the right time and the right place."

With the mission an enormous success, the company returned to Camp Bastion before departing to a known Taliban location on 5 June. The infantrymen patrolled the area for two days without encountering any enemy fire or IEDs.

"Between the two operations, we've kept the insurgency on their toes," Grassl said. "They don't know why we keep coming up there and harassing them, but we're harassing them to the point where they don't feel safe where they live, so they're on edge."

Note

Leatherneck 97, no. 8 (August 2014): 15–16. Reprinted with permission.

About the Author

Corporal Joseph Scanlan is a combat correspondent in Marine Expeditionary Brigade-Afghanistan (Task Force Leatherneck).

President Obama visits troops in Afghanistan on 26 May 2014 at Bagram Airfield. Obama thanked the soldiers, sailors, airmen, and Marines for their service. After his speech, he shook hands with every member present.

Official U.S. Air Force photo, courtesy of SSgt Evelyn Chavez

Statement by the President on Afghanistan

by President Barack H. Obama

The White House
Office of the Press Secretary
For Immediate Release
27 May 2014

Rose Garden
2:46 P.M. EDT

THE PRESIDENT: Good afternoon, everybody. As you know, this weekend, I traveled to Afghanistan to thank our men and women in uniform and our deployed civilians on behalf of a grateful nation for the extraordinary sacrifices they make on behalf of our security. I was also able to meet with our commanding general and ambassador to review the progress that we've made. And today, I'd like to update the American people on the way forward in Afghanistan and how, this year, we will bring America's longest war to a responsible end.

The United States did not seek this fight. We went into Afghanistan out of necessity, after our nation was attacked by al-Qaeda on 11 September 2001. We went to war against al-Qaeda and its extremist allies with the strong support of the American people and their representatives in Congress; with the international community and our NATO allies; and with the Afghan people, who welcomed the opportunity of a life free from the dark tyranny of extremism.

Defense Imagery 1631618, photo by CWO3 Benjamin Barr

Marines with Regional Command (Southwest) board a KC-130 Hercules aircraft to depart the Bastion-Leatherneck Complex in Helmand Province on 27 October 2014. The Marines departed following the end of the command's operations in Helmand.

We have now been in Afghanistan longer than many Americans expected. But make no mistake—thanks to the skill and sacrifice of our troops, diplomats, and intelligence professionals, we have struck significant blows against al-Qaeda's leadership, we have eliminated Osama bin Laden, and we have prevented Afghanistan from being used to launch attacks against our homeland. We have also supported the Afghan people as they continue the hard work of building a democracy. We've extended more opportunities to their people, including women and girls. And we've helped train and equip their own security forces.

Now, we're finishing the job we started. Over the last several years, we've worked to transition security responsibilities to the Afghans. One year ago, Afghan forces assumed the lead for combat operations. Since then, they've continued to grow in size and in strength, while making huge sacrifices for their country. This transition has allowed us to steadily draw down our own forces—from a peak of 100,000 U.S. troops, to roughly 32,000 today.

[Thus] 2014, therefore, is a pivotal year. Together with our allies and the Afghan government, we have agreed that this is the year we will conclude our combat mission in Afghanistan. This is also a year of political transition in Afghanistan. Earlier this spring, Afghans turned out in the millions to vote in the first round of their presidential election—defying threats in order to determine their own destiny. And in just over two weeks, they will vote for their next president, and Afghanistan will see its first democratic transfer of power in history.

In the context of this progress, having consulted

Defense Imagery 1173510, photo by Sgt Frances Johnson

In one of their final visits to the area of operations, Commandant Gen James F. Amos and SgtMaj Micheal P. Barrett, sergeant major of the Marine Corps, speak with Marines and sailors aboard Camp Leatherneck.

with Congress and my national security team, I've determined the nature of the commitment that America is prepared to make beyond 2014. Our objectives are clear: disrupting threats posed by al-Qaeda; supporting Afghan security forces; and giving the Afghan people the opportunity to succeed as they stand on their own.

Here's how we will pursue those objectives. First, America's combat mission will be over by the end of this year. Starting next year, Afghans will be fully responsible for securing their country. American personnel will be in an advisory role. We will no longer patrol Afghan cities or towns, mountains or valleys. That is a task for the Afghan people.

Second, I've made it clear that we're open to cooperating with Afghans on two narrow missions after 2014: training Afghan forces and supporting counterterrorism operations against the remnants of al-Qaeda.

Today, I want to be clear about how the United States is prepared to advance those missions. At the beginning of 2015, we will have approximately 98,000 U.S.— Let me start that over, just because I want to make sure we don't get this written wrong. At the beginning of 2015, we will have approximately 9,800 U.S. servicemembers in different parts of the country, together with our NATO allies and other partners. By the end of 2015, we will have reduced that presence by roughly half, and we will have consolidated our troops in Kabul and on Bagram Airfield. One year later, by the end of 2016, our military will draw down to a normal embassy presence in Kabul, with a security assistance component, just as we've done in Iraq.

Now, even as our troops come home, the international community will continue to support Afghans as they build their country for years to come. But our relationship will not be defined by war—it will be shaped by our financial and development assistance, as well as our diplomatic support. Our commitment to Afghanistan is rooted in the strategic partnership that we agreed to in 2012. And this plan remains consistent with discussions we've had with our NATO allies. Just as our allies have been with us every step of the way in Afghanistan, we expect that our allies will be with us going forward.

Third, we will only sustain this military presence after 2014 if the Afghan government signs the bilateral security agreement that our two governments have already negotiated. This agreement is essential to give our troops the authorit[y] they need to fulfill their mission, while respecting Afghan sovereignty. The two final Afghan candidates in the run-off election for president have each indicated that they would sign this agreement promptly after taking office. So I'm hopeful that we can get this done.

Defense Imagery 1629687, photo by SSgt John Jackson

Marines from 1st Battalion, 2d Marines, prepare to load onto a KC-130 aircraft on the Camp Bastion flightline on 27 October 2014. The battalion was the final infantry battalion to serve in Helmand Province.

The bottom line is, it's time to turn the page on more than a decade in which so much of our foreign policy was focused on the wars in Afghanistan and Iraq. When I took office, we had nearly 180,000 troops in harm's way. By the end of this year, we will have less than 10,000. In addition to bringing our troops home, this new chapter in American foreign policy will allow us to redirect some of the resources saved by ending these wars to respond more nimbly to the changing threat of terrorism, while addressing a broader set of priorities around the globe.

I think Americans have learned that it's harder to end wars than it is to begin them. Yet this is how wars end in the twenty-first century—not through signing ceremonies, but through decisive blows against our adversaries, transitions to elected governments, security forces who take the lead and ultimately full responsibility. We remain committed to a sovereign, secure, stable, and unified Afghanistan. And toward that end, we will continue to support Afghan-led efforts to promote peace in their country through reconciliation. We have to recognize that Afghanistan will not be a perfect place, and it is not America's responsibility to make it one. The future of Afghanistan must be decided by Afghans. But what the United States can do—what we will do—is secure our interests and help give the Afghans a chance, an opportunity to seek a long, overdue and hard-earned peace.

America will always keep our commitments to friends and partners who step up, and we will never waver in our determination to deny al-Qaeda the safe haven that they had before 9/11. That commitment is embodied by the men and women in and out of uniform who serve in Afghanistan today and who have served in the past. In their eyes, I see the character that sustains American security and our leadership abroad. These are mostly young people who did not hesitate to volunteer in a time of war. And as many of them begin to transition to civilian life, we will keep the promise we make to them and to all veterans, and make sure they get the care and benefits that they have earned and deserve.

This 9/11 generation is part of an unbroken line of heroes who give up the comfort of the familiar to serve a half a world away, to protect their families and communities back home, and to give people they never thought they'd meet the chance to live a better life. It's an extraordinary sacrifice for them and for their families. But we shouldn't be surprised that they're willing to make it. That's who we are as Americans. That's what we do.

Tomorrow, I will travel to [U.S. Military Academy] West Point and speak to America's newest class of military officers to discuss how Afghanistan fits into our broader strategy going forward. And I'm confident that if we carry out this approach, we can not only responsibly end our war in Afghanistan and achieve the objectives that took us to war in the first place, we'll also be able to begin a new chapter in the story of American leadership around the world.

Thanks very much.

END
2:58 P.M. EDT

Cpl Jeffery Mount, SFAAT 2-215, prepares to depart FOB Nolay. SFAAT 2-215 is completing a seven-month deployment as the last advisors in support of the 2d Brigade, 215th Corps, ANA.

Defense Imagery 1314844, photo by Cpl Joshua Young

Last Marines Exit Sangin, Afghanistan

by Corporal Joshua Young

Leatherneck, July 2014

FOB Nolay, Afghanistan—The last Marines at Forward Operating Base (FOB) Nolay exited the Sangin Valley, leaving the 2d Brigade, 215th Corps, Afghan National Army (ANA) in full control of the FOB and the surrounding area.

Marines from Security Force Assistance Advisor Team (SFAAT) 2-215 arrived at Camp Leatherneck following a seven-month deployment in a convoy out of Sangin Valley [on] 5 May. On the same day, the last leathernecks of 1st Battalion, 7th Marine Regiment, departed FOB Sabit Qadam, also in Sangin.

The SFAAT 2-215 Marines and corpsmen were part of the last U.S. brigade-level advisor team in Regional Command Southwest, which includes Helmand and Nimroz Provinces. (The British continue to advise the 3d Brigade in central Helmand.)

Marines preparing to redeploy expressed their feelings on leaving the FOB completely in the hands of 2d Brigade. "The 2d Brigade has had the torch for some time," said Captain Joseph Dewson, an advisor with SFAAT 2-215. "We have been able to spend time with them and vested a personal interest in their success."

Colonel Christopher Douglas, the team leader for SFAAT 2-215, kept a "no interference" stance with his advisory unit. One of his goals was to rapidly allow the brigade to stand on its own with as little advisory help as possible to ensure its capability to defend the Sangin Valley from hostile attacks.

By adopting that stance, the 2d Brigade has developed sustainable "Afghan solutions to Afghan problems" and, in turn, has become more confident in its problem-solving ability and ability to carry on the mission without coalition assistance. Many of the advisors believe the Afghan solutions work better than some of the Coalition force solutions used in the past. Although the "no interference" stance was kept, the advisors gave feedback to the leaders of 2d Brigade at every opportunity to reinforce their actions as well as to provide a positive learning environment.

"I feel honored to have been part of the legacy left by previous Marines and Coalition forces," said Douglas. "I expect to see them continue to focus on stability and actively opposing insurgent violence, to dominate and win every fight."

For some of the advisors, this was not their first time at FOB Nolay. Gunnery Sergeant John W. Greene, who was shot in [the] shoulder while returning enemy fire during November 2013, conducted monthly visits to the base from February 2011 to February 2012 as the division utilities chief for 2d Marine Division. He expressed his feelings on finally leaving after his second deployment in the area.

"After being at FOB Nolay on my last deployment where it was all Marines, to now where the Afghans run the FOB, it feels good that the ANA are doing so well," said Greene. "The 2d Brigade has been running strong, planning hard and executing missions on their own for a while now. They're doing a super job denying the insurgents free movement of control."

Corporal Robert Santiago, the squad leader of FOB Nolay's security force, nicknamed "The Guardians of Nolay," worked to ensure the base was secure at all times. Although his role on the base did not allow him to directly advise the Afghans, he and his Marines always were present to provide security for the advisors. "It feels good to see a positive result from all our efforts here," said Santiago. "I'm expecting to see more of the same great work they've been completing the entire time we've been working with them."

Major Paul D. Tremblay was the commander of Bravo Company, 1st Battalion, 6th Marines, on FOB Nolay in 2011. He was involved in the major clearing operation that secured the upper Sangin Valley and the Kajaki District. This deployment's end marks Tremblay's second full deployment to FOB Nolay.

"I couldn't be more humbled to have the opportunity to be lucky enough to find myself in this position to close the doors and shut off the lights on such a prestigious base in Helmand Province," said Tremblay, the deputy team leader, SFAAT 2-215. "We've definitely set a foundation that Afghanistan can take the rest of the way forward."

The Sangin Valley is a hotbed for nefarious and illegal activities. It is strategic in its proximity to major corridors, such as Route 1 and Route 611.

Drug runners and insurgents often travel on Route 1, which runs all the way through Afghanistan from Pakistan to Iran. The two routes are a crossroads for both trade and drug trafficking. Much of the Taliban's funding comes from the profits of the poppy harvests. Black tar heroin is extracted from the poppy plants, and the drugs are shipped all over the world.

The Taliban controls much of the heroin trade and are dependent on the industry. When the weather cools off, the insurgency turns toward facilitating the poppy planting. When planting begins, fighting almost ceases.

In 2006, when they first entered the Sangin Valley after the resurgence of the Taliban, the

Defense Imagery 1164084, photo by Cpl Joshua Young

SFAAT 2-215 provides security during a medical evacuation on an ANA base near FOB Nolay, Helmand Province.

Coalition forces had the lead role in all combat operations. During the course of the campaign, the lead has steadily been turned over to Afghan forces as the Coalition took on an advisory role.

Success in the region has not come easily. Many from Coalition forces and the Afghan National Security Forces have paid the ultimate price to bring stability to the war-torn area.

One of the most profound examples that illustrate the 2d Brigade's eagerness to continue the fight and completely make it the unit's own is the changing of the base name. Forward Operating Base Nolay now is known as Forward Operating Base Hamidullah.

It was named after Lieutenant Hamidullah, who was a platoon commander in the 2d Brigade's reconnaissance tolay (company). He was known for multiple heroics over the years and as one of the best platoon commanders and fighters in the 215th Corps.

"As soon as the brigade's executive officer told me they'd decided to change the name, I got excited," Tremblay said. "It's proof that we've done everything we can to set the conditions for transition, the effectiveness on how the Afghans themselves have accepted the transition and are willing to take ownership of it, make it their own and take it to the next step and the rest of the way.

"The Afghans of the 2d Brigade have chosen to name these positions after their heroes and warfighters who had a reputation above and beyond their individual sacrifice on any particular day," he continued. "Guys who we would hold up on high as an example of honor, courage and commitment."

Forward Operating Base Robinson also received a name change.* It now is known as FOB Rahatullah, named after Major Rahatullah, who was killed in an operation to disrupt enemy forces in support of the 5 April presidential elections.

Note

Leatherneck 97, no. 7 (July 2014): 12–13. Reprinted with permission. Copyright *Leatherneck*.

About the Author

Corporal Joshua Young was assigned as a combat correspondent with Marine Expeditionary Brigade-Afghanistan (Task Force Leatherneck).

---

* FOB Robinson was originally named for SSgt Christopher L. Robinson with 2d Battalion, 20th Special Forces Group, who was killed in action on 25 March 2006.

Part VI
Looking Back

Setting Up a Patrol Base, Nawa District by Michael D. Fay.
Marine Corps Art Collection

From 2010 to 2014, the Marine Corps approached its increased role in Afghanistan with enthusiasm, employing the Corps' traditional expertise in counterinsurgency (COIN) warfare. Additionally, the newly created U.S. Marine Corps Forces, Special Operations Command (MARSOC) fully participated in U.S. Special Operations Command's Afghanistan missions during this period.

After years of effort, the Corps' Camp Leatherneck stood as the last American base in Afghanistan. Marines turned the base over to the Afghan military in October 2014. Unfortunately, this act proved premature. Since 2014, the insurgency in Afghanistan has increased, and the U.S. Army has redeployed major units there. Marines remain in Afghanistan, serving in joint operations billets for the training mission. However, the Marine chapters of the Afghanistan war have concluded as of this writing.

A MARSOC Marine on patrol in Helmand Province.

Official U.S. Marine Corps photo, courtesy of Cpl Kyle McNally

MARSOC in Afghanistan, 2010–14

by Dr. Frank L. Kalesnik

From the beginning of the Global War on Terrorism in 2001, special operations forces played an increasingly crucial role in military operations across the globe. The U.S. Special Operations Command (USSOCOM) included Air Force, Army, and Navy Special Operations Forces (SOF) that performed a variety of missions, including civil affairs (CA), combating terrorism (CBT), direct action (DA), foreign internal defense (FID), information operations (IO), psychological operations (PSYOP), special reconnaissance (SR), and unconventional warfare (UW). From the beginning of Operation Enduring Freedom–Afghanistan (OEF-A), SOF contributed greatly to the success of the United States and its Coalition partners.

The Marine Corps began contributing combatant forces to Special Operations Command with the establishment of the U.S. Marine Corps Special Operations Command Detachment, or Det One, in 2003. Deployed to Iraq in 2004 as part of the Naval Special Warfare Squadron, Det One performed several raids and provided support to both conventional and other special operations units. While not a direct lineal predecessor of U.S. Marine Corps Forces, Special Operations Command (MARSOC), Det One, did help pave the way for the Marine command's establishment (at Secretary of Defense Donald H. Rumsfeld's insistence) in 2005. MARSOC was initially composed of a headquarters element and the Foreign

Defense Imagery 778665, photo by Cpl Kyle McNally

A MARSOC Marine waits in ambush for insurgents during a joint patrol with ANA Special Forces soldiers in Nahr-e Saraj District, Helmand Province, on 3 April. Marines with MARSOC's 1st MSOB recently returned from a nine-month deployment to Afghanistan, where they commanded the Special Operations Task Force-West and oversaw one of Afghanistan's bloodiest regions.

Military Training Unit (FMTU), later known as the Marine Special Operations Advisor Group (MSOAG). These Special Operations Marines began deploying for foreign internal defense (FID) missions almost immediately, providing training and advisory support to friendly military forces in countries around the globe.

Marine Special Operations Command quickly added 1st and 2d Marine Special Operations Battalions (MSOBs), which were formed by the transfer of 1st and 2d Force Reconnaissance Companies. With these additional forces, the Marine command prepared to deploy Marine special operations companies (MSOCs), initially with Marine expeditionary units (MEUs). The first company deployed with the 26th MEU in January 2007, subsequently detaching to support OEF-A. These companies remained with the Marine expeditionary unit until arriving in theater, at which point they came under the control of the theater special operations commander (TSOC) and detached from the unit, which they later rejoined for the return voyage home. Major General Mastin M. Robeson assumed command of MARSOC from Major General Dennis J. Hejlik on 24 July 2008, and soon decided to keep the companies then deployed in Afghanistan in theater, effectively ending the deployment of companies with expeditionary units. He explained his reasons in a 2009 oral history interview:

> [T]he minute the MEU would break the CENTCOM (U.S. Central Command)

Defense Imagery 778653, photo by Cpl Kyle McNally
A MARSOC Marine provides security at a landing zone in Nahr-e Saraj District, Helmand Province, on 28 March.

> waters, the SOCCENT (Special Operations Command Central) Commander would pull them into West Afghanistan, and then they would depart when the MEU departed; so there was a gap of anywhere from forty-five to ninety days between the companies [*sic*] presence. So in a COIN (counterinsurgency) fight, that's a disastrous way to do business.[1]

Major General Robeson's decision allowed companies "employed in Afghanistan . . . to properly turnover with the next MSOC and maintain operational continuity."

By February 2009, MARSOC established a two-company presence in Afghanistan, with 1st Marine Special Operations Battalion (MSOB) (located at Camp Pendleton, California) and 2d Marine Special Operations Battalion (located at Camp Lejeune, North Carolina) each providing companies as part of a rotating deployment cycle.

MSOCs included Marine special operations teams (MSOTs), which operated from distinct locations, usually forward operating bases (FOBs). They often partnered with friendly Afghan forces, providing training and advisory assistance that boosted their partners' operational capabilities. These were classic foreign internal defense missions, in which the 14-man teams worked to enhance the capabilities of larger allied units, acting as a combat multiplier that contributed significantly to operational success.

When the Marine Special Operations Regiment (MSOR) stood up in August 2009, the Marine command reorganized to establish three

Marine special operations battalions: two (1st and 2d MSOB) focused on OEF-A, while the 3d MSOB (formed from the MSOAG) provided teams for missions in other theaters (and sometimes Afghanistan). An important milestone was the deployment of 1st MSOB's command element as Special Operations Task Force 81 (SOTF-81), serving as a battalion-level headquarters for special operations forces in Afghanistan in November 2009. Major General Robeson recalled the circumstances that led to the MARSOC's deployment of its first special operations task force.

> I sat down with Admiral Olson (SOCOM Commander Admiral Eric T. Olson) in December of 2008, and said it's not that I can't do this, it's just that it's going to be very painful, and this has to be something that you, no kidding, think is that important for me to pull my command through this knot hole. And he looked me dead in the face and said, "Unless this breaks your Command, you need to do this. I need it and, to be candid, you need it more than you realize."[2]

By the end of 2012, three special operations task forces had deployed to Afghanistan. Upon assuming leadership of Special Operations Task Force-West (SOTF-W), the command element of the 1st MSOB was responsible for special operations in Badghis, Farah, Ghor, Helmand, Herat, and Nimroz Provinces. At this time, emphasis was placed on village stability operations (VSO) in which the special operations forces partnered with local Afghan police in an effort to provide security and foster self-governance at the village level. Of particular note was the pacification of the Upper Gereshk Valley in Helmand Province, an area described as "the most violent in all of Afghanistan." During this effort, SOTF-W coordinated the actions of an international force of 14,000 troops that included both conventional and special operations units. They also enabled Badghis Province to transition to Afghan governmental control, a significant achievement for a formerly volatile area.[3]

From 2010 until the conclusion of MARSOC's participation in OEF-A in October 2014, the command's 1st and 2d MSOBs continuously deployed Marines to Afghanistan as part of special operations task force staffs, with special operations companies, and as individual personnel serving in a variety of billets. The following list details the command's support of Operation Enduring Freedom–Afghanistan from 2010 to 2014.[4]

2010

1st Marine Special Operations Battalion

Marine Special Operations Task Force 81 HQ
(14 October 2009–31 May 2010)
Marine Special Operations Company 811
(25 July 2009–28 February 2010)
Marine Special Operations Company 812
(6 January 2010–31 July 2010)
Marine Special Operations Company 813
(26 August 2010–1 April 2011)

2d Marine Special Operations Battalion

Marine Special Operations Task Force-West HQ
(31 May 2010–1 February 2011)
Marine Special Operations Company 821
(18 November 2010–1 August 2010)
Marine Special Operations Company 822
(14 October 2009–3 June 2010)
Marine Special Operations Company 823
(30 April 2010–1 February 2011)

2011

1st Marine Special Operations Battalion

Marine Special Operations Company 811

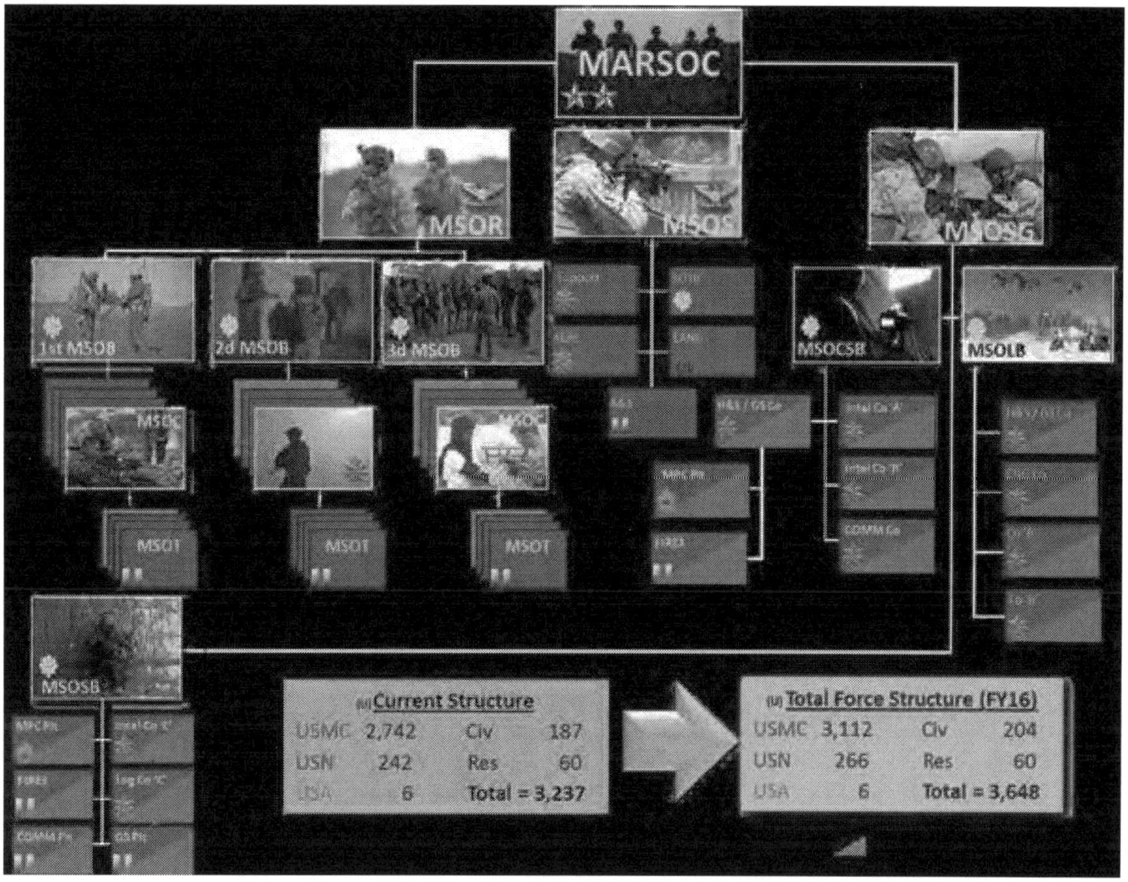

Marine Corps Forces, Special Operations Forces Command Brief, 2013

MARSOC organization.

(10 March 2011–30 October 2011)
Marine Special Operations Company 812
(29 September 2011–1 June 2012)
Marine Special Operations Company 813
(26 August 2010–April 2011)

2d Marine Special Operations Battalion
Marine Special Operations Task Force-West HQ
(31 May 2010–1 February 2011)
Marine Special Operations Company 821
(18 November 2010–1 August 2011)
Marine Special Operations Company 822
(12 June 2011–1 March 2012)
Marine Special Operations Company 823
(30 April 2010–1 February 2011)

3d Marine Special Operations Battalion
Marine Special Operations Team 8311 to Marine Special Operations Company 811
(29 June 2011–15 February 2012)

2012

1st Marine Special Operations Battalion
Special Operations Task Force 81.2
(4 January 2012–1 October 2012)
Marine Special Operations Company 811
(11 November 2012–7 July 2013)
Marine Special Operations Company 812
(29 September 2011–1 June 2012)
Marine Special Operations Company 813
(7 June 2012–6 February 2013)

2d Marine Special Operations Battalion
Special Operations Task Force 82.2 HQ
(1 October 2012–1 June 2013)
Marine Special Operations Company 821
(15 September 2012–15 April 2013)
Marine Special Operations Company 822
(12 June 2011–1 March 2012)
Marine Special Operations Company 823
(16 June 2012–15 August 2012)

3d Marine Special Operations Battalion
Marine Special Operations Team 8311 to Marine Special Operations Company 811
(29 June 2011–15 February 2012)
Marine Special Operations Team 8323 to Marine Special Operations Company 821
(13 August 2012–15 May 2013)

2013

1st Marine Special Operations Battalion
Marine Special Operations Company 811
(11 November 2012–7 July 2013)
Marine Special Operations Company 812
(7 July 2013–7 February 2014)
Marine Special Operations Company 813
(7 June 2012–6 February 2013)

2d Marine Special Operations Battalion
Special Operations Task Force 82.2
(1 October 2012–1 June 2013
Marine Special Operations Company 821
(15 September 2012–15 May 2013)
Marine Special Operations Company 822
(15 April 2013–15 November 2013)
Marine Special Operations Company 823
(7 November 2013–1 June 2014)

3d Marine Special Operations Battalion
Marine Special Operations Team 8323
(13 August 2012–15 May 2013)

2014

1st Marine Special Operations Battalion
Special Operations Task Force 81.3 HQ
(1 February–1 October 2014)

2d Marine Special Operations Battalion
Marine Special Operations Company 821
(16 May 2014–15 September 2014)
Marine Special Operations Company 823
(7 November 2013–1 June 2014)

MARSOC's efforts in Afghanistan had a significant impact on its growth. The Afghan experience shaped the assessment and selection of personnel, the evolution of doctrine, and training. The organization of the Marine special operations regiment, the Marine special operations support group, and their subordinate battalions also evolved significantly during this period. The following table of organization provides a snapshot of the command in 2013.

In an oral history interview conducted in 2015, MARSOC Deputy Commander Colonel Joseph Marello explained the impact of Afghanistan deployments on the command's maturation as a component Special Operations Command:

> If you go all the way back to the beginning, the concept was that first and second battalions would source what basically was a Maritime Special Force to the east and west coast MEUs. The FMTU, the Foreign Military Training Unit, which became the Marine Special Operations Advisory Group would do episodic JCETs (Joint Combined Exchange Training) and those kinds of missions. Capacity building, enabling kinds of missions. But it was Afghanistan and the requirements there that drove us to separate from the MEU.[5]

He described how the Afghan experience influenced the command's organizational structure:

> It also drove us to our reorganization, the first reorganization really, that got away from a company that looked like a Force Reconnaissance platoon, and a rifle platoon, and some enablers, to what we have today. An MSOC, with four of its teams, fully enabled. Now, that was a process that occurred over some time, but the first reorganization that established that foundation occurred, I wouldn't say it was because of Afghanistan, but just as we were making the movement from the MEU to deploying Marines directly to Afghanistan. The subsequent reorganization that followed was, I would say, almost entirely derived from lessons learned from Afghanistan. How to build an MSOT, an MSOC, how to enable it, and then how to build a Special Operations Task Force (SOTF), a battalion level headquarters task force.[6]

Colonel Marello explained the intensity of the command's Afghanistan commitment:

> At that time, in 2011 actually, the focus of effort for the command was sourcing forces to Afghanistan. It was almost all-consuming. At that time, we were maintaining a SOTF level headquarters of about 225 Marines, just for the headquarters, and then two fully enabled MSOCs. For the SOTF headquarters, we would do two battalion rotations of seven months each, and then the Army would pick up the third, and then we would come back again, so we would do two or three rotations.[7]

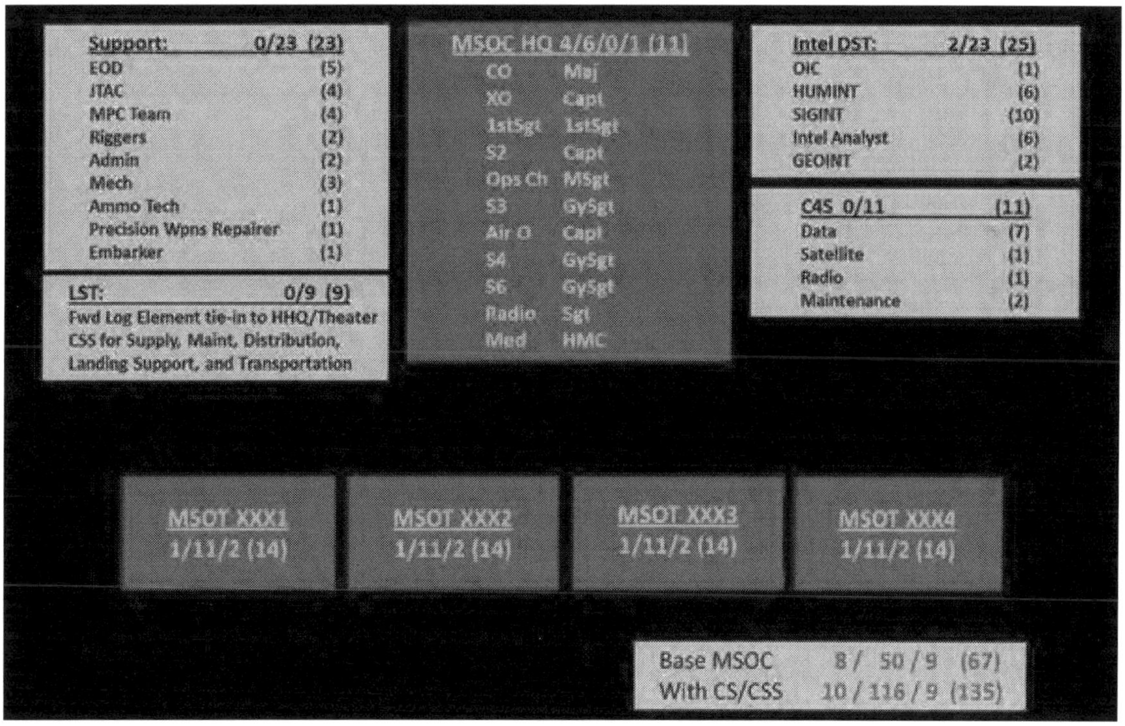

Marine Corps Forces, Special Operations Forces Command Brief, 2013

Marine Special Operations Company.

Defense Imagery 778648, photo by Cpl Kyle McNally

Coalition Special Operations Forces members provide security during a two-day presence patrol with Afghan commandos from the 9th Kandak in Farah Province on 27 February.

He observed, "It took just about all the capacity of the command to source that."[8] He also noted:

> And along with it, we didn't have the capacity to source all of the SOTF itself. We also relied on the Marine Corps for augmentation. We weren't fully built at that time. And so, going back to the Marine Corps at the quarterly synch conferences and confirming our requirements, and then following up through Headquarters Marine Corps for the actual sourcing, and then receiving and integrating a hundred plus Marines in each SOTF headquarters, was another sizable effort.[9]

It is important to recognize that, in addition to the critical skills operators from the Marine Special Operations Regiment who provided the "tip" of the MARSOC spear, the personnel of the Marine Special Operations Support Group and its subordinate battalions contributed the "enablers," or special operations capabilities specialists that gave the spear its shaft.* Operational requirements called for additional personnel from outside MARSOC. According to Colonel Marello:

> The SOTF headquarters required over 40 percent augmentation from big Marine Corps. So communicators, engineers, generator mechanics, intelligence analysts, you name it, the combat support/combat service support aspects of the SOTF headquarters required significant augmentation.[10]

The high operational tempo placed strains on

* The term *critical skills operators* refers to military occupational specialty 0372 and *special operations capabilities specialists* refers to military occupational specialty 8071.

the force. Former Marine Special Operations Support Group Commander Colonel Archibald M. McLellan explained:

> Intel guys, and communicators, and JTACS [joint terminal attack controllers], were getting run at a very cyclic, ragged rate. We did not have a good degree of predictability, because we were sending out MSOCs and we were sending out SOTFs, and everybody needed everything, and everything was a priority. So really, our guys were running at a less than a 1 to 2 dwell to support SOTFs and MSOCs as we were growing the force during the initial days. It was not uncommon to have waivers and not even think about deferring training, but actually waiving it just so we could get guys out there. When they'd come back from their first or second deployment, and they had some breathing space, why would you send them through the training after they'd had all this experience.[11]

Major General Paul E. Lefebvre served as the commanding general of Marine Special Operations Command from 2010 to 2012. In an interview published in *Marine Corps Outlook: 2010–2011 Edition*, he explained the SOTF's function:

> A key task for the SOTF is integration. On the battlefield we're on, with the other government agencies operating there—State Department, Afghan government and military, our partners from other nations—the SOTF integrates all those capabilities. SOTF means taking a Marine with a good understanding of counterinsurgency doctrine and putting him into the nebulous condition of ungoverned stations, where you want to understand human terrain. Marines are uniquely experienced in doing that from our OIF (Operation Iraqi Freedom) and OEF (Operation Enduring Freedom) operations. And that is the essence of what that SOTF is doing.[12]

The MSOCs subordinate to the task forces also performed a command and control function. The following chart depicts their table of organization. The base company (four teams and the headquarters element) were augmented with several support personnel. As the chart shows, a fully enabled company (with four teams, each of 1 Marine officer, 11 Marine enlisted, and 2 Navy corpsmen), could well exceed a hundred personnel with the addition of combat service and combat service support Marines.

In an interview with *The Year in Special Operations: 2012–2013 Edition*, Major Andrew Christian of MSOC A, 1st MSOB stated:

> When we deploy an MSOC, we have about 105 personnel. But SOCOM realized the MSOC can be the HQ [headquarters] for a much larger SOF (Special Operations Force), so we have taken the MSOC's headquarters and three teams and attached under that one Navy SEAL platoon and two to three ODAs [operational detatchment-alpha]. Additionally, we took a standard unit—[U.S. Army's] 1st [Battalion] of the 505th Parachute Infantry Regiment, 82nd Airborne [Division]. When you add all those attached forces, you have more than 500 soldiers, sailors, airmen, and Marines operating under one MSOC commander, which is a substantial capability.[13]

Major Christian went on to explain the companies' relationship with the task force:

> The MSOC, with those maneuver elements, has SOTF-West above us, a battalion-level

> unit responsible for the largest chunk of real estate in Afghanistan. SOTF-W also has a similar company operating further west, so it is a larger snapshot of what we were doing, with the mission to command all special forces in the western section of the country.[14]

MSOC activities during this period were an outstanding example of distributed operations. Though the 500-man formation described above seems large, it was operating across more than 50,000 square miles of western Afghanistan. As such, they were spread widely across key villages and engagement areas but were never isolated. They were able to maintain robust capabilities through the strong, multilayered command and control systems of the special operations task force and through the complete integration of critical enablers into combat formations all the way down to the team level. Each team was reinforced with at least one joint tactical air controller and a four-Marine intelligence team. The team was connected to national intelligence systems. The team could be further augmented with explosive ordinance disposal and military dog teams, all tied together through multiple data and voice communication systems. Each of these reinforced teams was capable of independent operations, producing intelligence for others as well as producing and acting on intelligence on their own area, while always cognizant that their "independent actions" were actually being coordinated and synchronized by the company and task force commanders to maximize their cumulative effects.[15]

The versatility of this task-organized special operations team was on display as special operations efforts turned to village stability operations (VSOs). Reinforced MSOTs were able to enter unsympathetic villages, establish a persistent presence there, and begin to turn the tide of popular opinion all the while leveraging critical combat support and combat service support capabilities to conduct the entire spectrum of SOF operations to kinetic F3EAD (find, fix, finish, exploit, analyze, and disseminate) operations to civil affairs and key leader engagements. Alone, the reinforced team was a potent multifunctional unit. Then, when paired with a partner force, the team helped the allied force take on significant combat operations with confidence and clear expectation of success.[16]

With MSOC's Afghan mission complete, the concept for its employment changed to one focused on subregional alignment. The MSOBs were redesignated as Raider battalions in 2015; now they each provide a company to theater special operation commands in Pacific Command (1st Raider Battalion), Central Command (2d Raider Battalion), and Africa Command (3d Raider Battalion). The companies are fully enabled to perform all of MSOC's core activities, and maintain a persistent presence in their assigned subregion. While special operations Marines are not currently deploying to Afghanistan, their experience there continues to influence their conduct of current operations, and will into the foreseeable future.[17]

Notes

1. MajGen Mastin M. Robeson, MARSOC commander, intvw conducted by MARSOC Command Historian Cynthia Hayden, 9 November 2009.
2. Ibid.
3. Cpl Kyle McNally, "SOTF-West Returns from Deployment," MARSOC Marine Corps, 30 November 2012.
4. Frank Kalesnik, MARSOC command historian, "MARSOC OEF-A Deployments, 2010–2014" (unpublished unit list, 22 September 2015).

5. Col Joseph Marello, oral history intvw, 2015.
6. Ibid.
7. Ibid.
8. Ibid.
9. Ibid.
10. Ibid.
11. Col Archibald M. McLellan, oral history intvw, 23 June 2015.
12. J. R. Wilson, "The State of MARSOC: Interview with MajGen Paul E. Lefebvre," Defense Media Network, 11 November 2010.
13. J. R. Wilson, "MARSOC Year in Review," *The Year in Special Operations: 2012–2013 Edition* (Tampa, FL: Defense Media Network, 2013): 36–42.
14. Ibid., 39–40.
15. Gary Oles, MARSOC Deputy G-3, comments to author, 25 August 2015.
16. Ibid.
17. "Concept of Employment" from Marine Corps Forces, Special Operations Forces Command Brief, 2013.

CWO2 James Law, the jump platoon commander and battalion gunner for 3d Battalion, 3d Marines, walks across a bridge in Nawa, Helmand Province, while providing security on 22 October 2010.

Defense Imagery 101022, photo by Sgt Mark Fayloga

Marine Commanders Reflect on 13 Years of War in Afghanistan

by Hope Hodge Seck

Marine Corps Times, 5 November 2014

It was late 2010, and Sergeant Philip A. McCulloch Jr. was locked into what may have been the darkest battle of America's longest war.

The gauntly thin 22-year-old squad leader for Kilo Company, 3d Battalion, 5th Marines, had gotten too used to seeing the men in his unit killed by roadside bombs or enemy fire in the Sangin Valley, a hotly contested district in Afghanistan's Helmand Province that buzzed with insurgent activity. He did not allow himself to feel grief; he just got angry.

That anger boiled over when Corporal Tevan L. Nguyen, 21 [years old], was killed by an improvised explosive device [on] 28 December [2010], shortly before the unit would return home. "Nguyen's death was the result of treachery," McCulloch said. Taliban representatives had brokered a temporary cease-fire with the unit, surrendering weapons and claiming they had cleared IEDs from the area where 3d Battalion, 5th Marines, patrolled.

When it became clear that the deal had been a lie, "it was game on," McCulloch said. "We made sure that we went out there and let them know they made a mistake. And they paid for it with their lives and a lot of blood, a lot."

Just over a week later, McCulloch, fueled by his emotion, would earn the Silver Star for exceptional valor in the face of death while leading an aggressive six-hour pursuit of the enemy that culminated in a dramatic air assault. When it was over, the battlefield was littered with dead Taliban fighters and their motorcycles.

Four years later, McCulloch still feels the loss of his brothers-in-arms. But he no longer believes the war they fought together was worth the cost they paid. As news broke in late October that the Marines had concluded their combat operations, McCulloch's skepticism and frustration regarding Afghanistan appears to be the rule rather than the exception. A half-dozen active-duty Marines who spoke with *Marine Corps Times* are unanimous in their belief that it is time for U.S. forces to pull out of the country after 13 years of combat. But most expressed doubt about the country's ability to sustain progress.

McCulloch stays in touch with his Afghan interpreter from Sangin, whom he says receives daily death threats now that the Marines have pulled out. Despite positive reports about the development, improvement, and fighting spirit of the Afghan National Army's 215th Corps, which took over for the Marines in Helmand Province, McCulloch said he has little faith in their mettle based on his own experiences working with them in 2010 and 2011. He feels the Marines' counterinsurgency strategy, with its village shuras, cash investments, and focus on building local trust and relationships made them less effective at defeating the enemy.

And above all, he believes that no amount of time would be sufficient to win a "thousand-year war" fueled by intricate Afghan tribal relationships and cultural codes. Before long, McCulloch said, it will be like the Marines were never there.

"I give it less than 12 months," he said.

A Marine sergeant who deployed to Afghanistan in 2012 as part of an embedded training team expressed faith in the Afghan troops' willingness and ability to fight the insurgency on their own, but said the U.S. forces were hopelessly overmatched when it came to effecting lasting change in a country beset by corruption and disarray. "How can 14 years fix 2,000 years?" he said.

Amid the pessimism and disappointment, former commanders of Marine forces in Afghanistan say that, while the future of Afghanistan is a mystery, the Marines have many reasons to be proud of their part in this long, hard war. Two things, specifically, they all agreed were certain: the Marines leave Afghanistan having achieved victory in their mission, and the Corps would never be the same for having served there.

Out of Afghanistan

Early in the morning on 26 October, the last 872 Marines remaining in Helmand Province boarded [Lockheed] C-130 [Hercules] jets and left for Kandahar Airfield, a brief stop on their way back to the United States. They left behind a few hundred million dollars worth of dusty desert tents and hard-shelled structures, all-terrain vehicles, and an airstrip. Little remained by way of monument to the tens of thousands of Marines who cycled through Camp Leatherneck since 2009, when President [Barack H.] Obama ordered a surge of 30,000 troops into Afghanistan. The neighboring Nimroz and Farah Provinces, which had seen hard-fought battles between Marines and the insurgency in previous years, had long since been turned back over to the Afghans, their bases dismantled or handed off.

The first Marine force had swept into Afghanistan just weeks after the 11 September 2001 terrorist attacks on the World Trade Center buildings and Pentagon. Led by then-Brigadier General

James N. Mattis, who would eventually become a four-star commander of U.S. Central Command, Task Force 58, was more than 1,000-strong, composed of the 15th Marine Expeditionary Unit [15th MEU], 26th MEU, and other Navy and Marine Corps elements.

From ships at sea, the task force would move inland on CH-53s to establish Forward Operating Base Rhino, southwest of Kandahar in the Registan Desert. They also took control of the international airport in Kandahar, assisted with security at the U.S. embassy in Kabul, and conducted a variety of reconnaissance missions.

But while the Marine Corps was first on the ground in 2001, its work in Afghanistan would begin in earnest some seven years later when the 24th Marine Expeditionary Unit returned to Kandahar, then moved into the Taliban stronghold of Garmsir in Helmand Province, a little-populated hotbed for insurgent activity in the southern reaches of Afghanistan, where poppy farming fueled a thriving drug trade. Later the same year, 2d Battalion, 7th Marines, deployed to Helmand and Farah Provinces, waging a pitched battle with insurgents from Now Zad to Delaram. When the surge hit its stride the following year, 11,000 Marines joined British troops operating in Helmand. As the surge troops established themselves in Helmand and southern Afghanistan, this region would become affectionately known as "Marineistan."

"I think we understood going into Afghanistan, when we got that mission, it was really the two-fisted fighter," said Major General Larry D. Nicholson, who as a one-star general in 2009 led the 10,000 troops who fell under Marine Expeditionary Brigade-Afghanistan. "We were going to go after the Taliban, but we were really going to go after the population." To that end, Nicholson said, the work was complex.

"Where we went, we cleared. Where we cleared, we stayed. And where we stayed, we built," Nicholson said. "We were very determined to go to those population centers. It's a remote area, but those population centers, we worked very hard to get there and start building, not American trust, but [Afghan] trust in their government."

Marjah and Sangin

Despite the Marines' focus on civil affairs and the tenets of counterinsurgency, it became clear, said Nicholson, that they would have to aggressively pursue the insurgency deep into its own territory.

"We were reminded constantly, 'Yeah, the Marines are doing okay there [in Helmand]. We're impressed with what the British and Marines and Danes and Estonians are doing,' " Nicholson said. " 'But the enemy's in Marjah. When are you going to go there?' "

The urban region of Marjah, Nicholson said, had all but seceded from the rest of Helmand Province. "They flew a Taliban flag," he said. "And we took that away from them."

Launched 13 February 2010, Operation Moshtarak was led by 1,500 troops with 1st Battalion, 6th Marines, in conjunction with British, Afghan, and allied international troops. As the fighting stretched on for months, casualty rates climbed. Dozens of Marines would ultimately become casualties of the Marjah campaign. Progress in the region was slow as troops struggled to win the trust of local civilians they sought to help.

In May 2010, Army General Stanley A. McChrystal, then the commander of all allied forces in Afghanistan, called Marjah a "bleeding ulcer" on the war effort. The region would ultimately become one of Helmand Province's success stories, commanders who oversaw the effort there said, with Taliban activity quenched, schools and markets [were] built and restored, and recently a peaceful presidential election [was held].

But that would not take place, said Lieutenant General Richard P. Mills, former two-star commander of Regional Command-Southwest [RC-SW] overseeing Helmand, until the Marines launched what would be their most difficult and costly Afghan offensive yet.

"We took the pressure off Marjah by going into Sangin," said Mills, now the commanding general of Marine Forces Reserve. "Sangin was critical to the enemy; it was an area the enemy was not going to give up. And the minute we went into Sangin, the pressure in Marjah was relieved."

Ultimately, Mills said, sending Marines into Sangin was the right decision. But nothing about it was easy.

During the 2010 assault on Sangin, 25 members of 3d Battalion, 5th Marines, would lose their lives. The death toll in Sangin would ultimately rise to some 50 Marines and 100 British troops, who had held a presence in the Taliban hotbed and opium hub even before the Marines arrived. But when the Marines finally handed Sangin over to members of the 215th Corps this spring, it was on the heels of a relatively peaceful primary election. The insurgents, while still active, had been pushed to the fringes of the population centers, at least for the time being.

"That's the tough part of being a commander," Mills said. "You hope that the courage and bravery and dedication of those young Marines, that they can stick to it, and they did. But every casualty report is tough. They hurt, they hit home. As commander, you have to look at the strategic level of what you're trying to do."

Just a few years later, it was clear, Nicholson said, that these Afghanistan campaigns had an elite place in Marine Corps history.

"The iconic battles of this last decade have been . . . Fallujah, Ramadi, Sangin, and Marjah," he said. "We fought like hell to take those areas."

Lessons from Afghanistan

The Marine Corps emerges from Afghanistan a changed force, due in part to the maturity and ingenuity gained from 13 years at war and in part from the passage of time.

Afghanistan saw the first major combat mission for the [Bell Boeing] MV-22B Osprey tiltrotor aircraft, which now is an operational staple for the Marines in practically every theater imaginable. The Marines also benefited from the M142 High Mobility Artillery Rocket System [HIMARS], a truck-mounted long-range rocket they had used first in Anbar Province, Iraq, but employed more heavily in Helmand Province, deploying 400 of the powerful rounds by January 2013.

The war also saw the extended use of Marine Corps Forces Special Operations Command. MARSOC, which activated in 2006, matured throughout the war. The first deployment of a MARSOC company to Afghanistan in early 2007 was a disaster that resulted in as many as 19 civilian casualties and an order from the U.S. Army to return home early. But the rule breaking and heedlessness that allegedly marked that first deployment [were] soon quelled.

To date, MARSOC troops have earned five Navy Crosses for heroism in Afghanistan, [with] about a third of those awarded in that time period. Mills said MARSOC teams, operating in and around Helmand Province, were a force multiplier for the Marines.

"They were operating kind of semi-independently," he said. "They were able to cover down on an area and backstop local security forces. They showed the enemy we have the ability to operate wherever and whenever we wanted to. And they did a tremendous job in combat."

As the mission in Afghanistan progressed, the Marines also prioritized cultural training and basic language classes, briefing Marines on how to relate to Afghan civilians. This included sit-down classes and having troops attend shuras at mock-Afghan villages staffed by role-players.

Following a surge in deadly insider attacks targeting Marines in 2012 and 2013, the Corps implemented more advanced cultural and language integration for senior officers that drilled them on, among other things, the tenets of Afghan cultural norms. Marines in Afghanistan told *Marine Corps Times* earlier this year that they credited this training with defusing more potential insider attacks.

Nicholson said this cultural learning strategy had been critical to success in Marjah.

"We spent an inordinate amount of time trying to understand the human terrain in Marjah. We had PhDs helping us try to understand that human dynamic," he said. "The trick now, frankly, is, that we don't lose it. All these skill sets that we've acquired, don't let them get away."

Nicholson is now pushing to rewrite the Marines' *Small Wars Manual* to incorporate the specific cultural lessons learned in Afghanistan.

The Question of Victory

The war in Helmand Province has felt progressively less like a war since the Marines began transferring authority of various regions back to Afghan forces more than two years ago. In the last 12 months, even the Marines' advise-and-mentor mission has dwindled to nothing as officials encouraged the 215th Corps to operate independently. As troops stood guard outside Camp Leatherneck with few missions outside the base and little formal monitoring of what took place in the regions from which they had departed, it was a strange way to end a war.

Brigadier General Daniel D. Yoo, the commander of RC-SW and Marine Expeditionary Brigade-Afghanistan during its last year in Helmand, said this lack of action had proven to be one of the hardest things for Marines in the war's final phase.

"We're very aggressive by nature," he said. "We look for solutions, there's no doubt about it."

What had helped to counter this frustration, he said, was observing the progress the 215th Corps had made since it was created from scratch in 2009 with a few handpicked leaders from the 205th Corps in Kandahar. The Marines helped Afghan troops create and man a regional corps battle school that equips new soldiers with a higher level of standardized training in Helmand, so the Corps does not have to send them to Kabul and risk losing them to the bigger city. They also trained Afghan tactical air controllers for the Corps' new fleet of [Mil] Mi-17 [Hip-H transport] and Mi-35 [Hind E gunship] helicopters.

A visit to the battle school's firing range makes clear that some of the Afghans' military skills remain rudimentary, but Yoo points to Sangin as evidence of what their future could hold. Following the Marines' departure, an intense fighting season resulted in heavy ANA [Afghan National Army] and civilian casualties. Afghan-held checkpoints changed hands, but Marine officials said the troops were ultimately able to regain them all. Yoo said a successful outcome in contested regions like Sangin might be a brokered peace or arranged coexistence between the government and Taliban and tribal factions.

"As long as the district centers are open, as long as the major lines of communication are open in order to provide services, and people can move from district center to district center, that to me is somewhat stable," Yoo said. "It's when, what we saw in the past when the insurgents ruled everything to include the local law, that's when I think it's probably a perspective that there's failure there."

Defense Imagery 1631686, photo by Sgt Dustin D. March

BrigGen Daniel D. Yoo (left) and SgtMaj Douglas E. Berry (right) walk toward a hangar after arriving at Kandahar Airfield, Afghanistan, on 27 October 2014. The Marines transitioned to Kandahar after the end of RC-SW operations in Helmand Province.

Did the Marines Achieve Victory in Their Helmand Mission?

"Absolutely," Yoo said.

Like Yoo, Mills said time will prove the long-term effect of the Marines' efforts and the Afghan troops' ability to sustain the change. "I expect the insurgents to test the force that we left behind; I think that everybody expects that," he said. "My prediction is that will surge over the next fighting season, and that will be the ultimate decision. The next two or three years will be critical." He argued that accomplishments, such as building schools, viewed by some Marines to be extraneous to the task of warfighting, were actually crucial to the success of Helmand Province. "It showed that the insurgents could no longer threaten the people into doing what they wanted. That was an important metric on our success. Because the Taliban hated education."

Nicholson said he was cautiously optimistic about the future of Helmand, though he said it was imperfect and a work in progress. He was proud, he said, of what the Marines had done. For the country of Afghanistan, he said, the proof of victory is in peace. "A country at peace with itself and its neighbors, able to provide for its own security, that kind of smells like victory to me," he said.

Mattis, the legendary Marine commander who led the invasion of Afghanistan, declined to speculate on the long-term effects of the Marines' work there, saying he had a variety of reasons for wanting to remain silent on the topic for now. But he offered a statement of appreciation for the work of the Marines nonetheless.

"General Mattis might have said the Marines are no better friend and no worse enemy," he said, "but it was the lads in the infantry and the engineers who proved it beyond a doubt."

Note

Marine Corps Times, 5 November 2014. Reprinted with permission.

About the Author

Hope Hodge Seck was a staff writer with the *Marine Corps Times* from 2013 to 2015.

Appendices

Appendix A:
Command and Staff List

2010

International Security Assistance Force, Regional Command-Southwest
[ISAF RC-SW]

2d Marine Expeditionary Brigade [2d MEB]

Commanding General: BGen Lawrence D. Nicholson
Deputy: Col George S. Amland
Chief of Staff: Col William P. McLaughlin
Sergeant Major: SgtMaj Ernest K. Hoopii

G-1: LtCol John W. Bicknell Jr.
G-2: LtCol Scott T. Derkach
G-3: Col Michael P. Killion
G-4: Col Christopher B. Edwards
G-6: Col Allan M. Faxon Jr.

Command Element

2d MEB Headquarters Group
Commanding Officer: LtCol Christopher L. Naler

Marine Ground Combat Element

Regimental Combat Team 2 [RCT-2]
Commanding Officer: Col P. J. Kennedy

1st Battalion, 2d Marines [1st Bn, 2d Mar]
Commanding Officer: LtCol Michael A. Manning

Regimental Combat Team 7 [RCT-7]
Commanding Officer: Col Randall P. Newman

2d Battalion, 2d Marines [2d Bn, 2d Mar]
Commanding Officer: LtCol John E. McDonough

1st Battalion, 3d Marines [1st Bn, 3d Mar]
Commanding Officer: LtCol James M. Baker

3d Battalion, 4th Marines [3d Bn, 4th Mar]
Commanding Officer: LtCol Martin F. Wetterauer III

1st Battalion, 6th Marines [1st Bn, 6th Mar]
Commanding Officer: LtCol Calvin A. Worth

3d Battalion, 6th Marines [3d Bn, 6th Mar]
Commanding Officer: LtCol Brian S. Christmas

3d Battalion, 10th Marines [3d Bn, 10th Mar]
Commanding Officer: LtCol Todd R. Finley

4th Light Armored Reconnaissance Battalion [4th LAR]
Commanding Officer: LtCol Michael S. Martin

Marine Aviation Combat Element

Marine Aircraft Group 40 [MAG-40]
Commanding Officer: Col Kevin S. Vest

Marine Logistics Combat Element

Combat Logistics Regiment 2 [CLR-2]
Commanding Officer: Col John W. Simmons

I Marine Expeditionary Force (Forward) [I MEF (Fwd)]

Commanding General: MajGen Richard P. Mills
Deputy: BGen George P. R. Norton, United Kingdom-Afghanistan [UK-A]
Chief of Staff: Col Kevin F. Frederick
Sergeant Major: SgtMaj Micheal P. Barrett

G-1: LtCol Edward A. Garland
G-2: Col Steven M. Hanson
G-3: Col Barry J. Fitzpatrick Jr.
G-4: Col John C. Malik
G-6: Col Nicholas S. Chapman, UK-A

Command Element

2d Radio Battalion (Forward) [2d Radio Bn (Fwd)]
Commanding Officer: LtCol Matthew G. Rau

8th Communication Battalion (Forward) [8th Comm Bn (Fwd)]
Commanding Officer: LtCol Michael B. Schweighardt

2d Air Naval Gunfire Liaison Company [2d ANGLICO]
Commanding Officer: LtCol Leland W. Suttee

Marine Ground Combat Element

Regimental Combat Team 1 [RCT-1]
Commanding Officer: Col David J. Furness

2d Battalion, 1st Marines [2d Bn, 1st Mar]
Commanding Officer: LtCol Matthew S. Reid

3d Battalion, 1st Marines [3d Bn, 1st Mar]
Commanding Officer: LtCol Benjamin T. Watson

2d Battalion, 6th Marines [2d Bn, 6th Mar]
Commanding Officer: LtCol Kyle B. Ellison

2d Battalion, 3d Marines [2d Bn, 3d Mar]
Commanding Officer: LtCol John W. Evans Jr.

Regimental Combat Team 2 [RCT-2]
Commanding Officer: Col P. J. Kennedy

1st Battalion, 2d Marines [1st Bn, 2d Mar]
Commanding Officer: LtCol Michael A. Manning

3d Battalion, 3d Marines [3d Bn, 3d Mar]
Commanding Officer: LtCol Jeffrey C. Holt

3d Battalion, 6th Marines [3d Bn, 6th Mar]
Commanding Officer: LtCol Brian S. Christmas

2d Battalion, 9th Marines [2d Bn, 9th Mar]
Commanding Officer: LtCol James R. Fullwood

Regimental Combat Team 7 [RCT-7]
Commanding Officer: Col Randall P. Newman

1st Battalion, 3d Marines [1st Bn, 3d Mar]
Commanding Officer: LtCol James M. Baker

3d Battalion, 7th Marines [3d Bn, 7th Mar]
Commanding Officer: LtCol Clay C. Tipton

1st Light Armored Reconnaissance Battalion [1st LAR]
Commanding Officer: LtCol Scott D. Leonard

4th Light Armored Reconnaissance Battalion [4th LAR]
Commanding Officer: LtCol Michael S. Martin

1st Reconnaissance Battalion [1st Recon]
Commanding Officer: LtCol Michael J. Mooney

Marine Aviation Combat Element

3d Marine Aircraft Wing (Forward) [3d MAW (Fwd)]
Commanding General: BGen Andrew W. O'Donnell Jr.
Assistant Commander: Col Eric K. Fippinger
Chief of Staff: Col Shaun L. Sadler
Sergeant Major: SgtMaj Anthony A. Spadaro

G-1: Maj Kristin L. McCann
G-2: LtCol George J. David
G-3: Col Jack P. Monroe
G-4: LtCol James A. Coler
G-6: LtCol Robert K. Maldonado

Marine Air Control Group 38 (Forward) [MACG-38 (Fwd)]
Commanding Officer: LtCol Thomas J. Dodds

Marine Heavy Helicopter Squadron 361 [HMH-361]
Commanding Officer: LtCol Douglas V. Glasgow

Marine Heavy Helicopter Squadron 362 [HMH-362]
Commanding Officer: LtCol Thomas A. Pecina

Marine Light Attack Helicopter Squadron 169 [HMLA-169]
Commanding Officer: LtCol Brendan Reilly

Marine Fighter Attack Squadron 122 [VMFA-122]
Commanding Officer: LtCol John A. Bolt

Marine Aerial Refueler Transport Squadron 252 (-) [VMGR-252 (-)]
Commanding Officer: Maj Brent A. Johnson

Marine Medium Tiltrotor Squadron 264 [VMM-264]
Commanding Officer: LtCol Jeffrey P. Hogan

Marine Medium Tiltrotor Squadron 365 [VMM-365]
Commanding Officer: LtCol Craig C. LeFlore

Marine Unmanned Aerial Vehicle Squadron 2 [VMU-2]
Commanding Officer: LtCol Daniel T. Lathrop

Marine Aviation Logistics Squadron 16 [MALS-16]
Commanding Officer: LtCol James C. Carroll III

Marine Wing Support Squadron 373 [MWSS-373]
Commanding Officer: LtCol Thomas M. Fahy

2011

9th Air and Space Expeditionary Task Force-Afghanistan [9th AETF-A]

Marine Tactical Electronic Warfare Squadron 1 [VMAQ-1]
Commanding Officer: LtCol Chandler P. Seagraves

Marine Tactical Electronic Warfare Squadron 3 [VMAQ-3]
Commanding Officer: LtCol Marty A. Moore

Marine Tactical Electronic Warfare Squadron 4 [VMAQ-4]
Commanding Officer: LtCol Marlin C. Benton

International Security Assistance Force, Regional Command-Southwest [ISAF RC-SW]

I Marine Expeditionary Force (Forward) [I MEF (Fwd)]

(January–March 2011)

Commanding General: MajGen Richard P. Mills
Deputy: BGen George P. R. Norton, UK-A
Chief of Staff: Col Kevin F. Frederick
Sergeant Major: SgtMaj Micheal P. Barrett

G-1: LtCol Edward A. Garland
G-2: Col Steven M. Hanson
G-3: Col Barry J. Fitzpatrick
G-4: Col John C. Malik
G-6: Col Nicholas S. Chapman, UK-A

Command Element

2d Radio Battalion (Forward) [2d Radio Bn (Fwd)]
Commanding Officer: LtCol Matthew G. Rau

8th Communication Battalion (Forward) [8th Comm Bn (Fwd)]
Commanding Officer: LtCol Michael B. Schweighardt

2d Air Naval Gunfire Liaison Company [2d ANGLICO]
Commanding Officer: LtCol Leland W. Suttee

Marine Ground Combat Element

1st Marine Division (Forward) [1st MarDiv (Fwd)]
Commanding General: BGen Joseph L. Osterman
Chief of Staff: Col Robert F. Castellvi
Sergeant Major: SgtMaj Phillip A. Fascetti

G-1: Maj Daniel C. Hench (until March 2011)
LtCol Keither R. Moore
G-2: LtCol Jeffrey H. Hauser
G-3: Col Robert L. Gardner
G-4: Col William B. Spahn
G-6: LtCol Richard J. Bordonaro

Regimental Combat Team 1 [RCT-1]
Commanding Officer: Col David J. Furness
Executive Officer: LtCol Robert K. Schwarz
Sergeant Major: SgtMaj Bryan K. Zickefoose

Regimental Combat Team 2 [RCT-2]
Commanding Officer: Col P. J. Kennedy
Executive Officer: LtCol Steven J. Grass
Sergeant Major: SgtMaj G. W. Young

2d Battalion, 1st Marines [2d Bn, 1st Mar]
Commanding Officer: LtCol Matthew S. Reid

2d Battalion, 3d Marines [2d Bn, 3d Mar]
Commanding Officer: LtCol Christopher G. Dixon (until December 2011)
LtCol Charles R. Cassidy

3d Battalion, 5th Marines [3d Bn, 5th Mar]
Commanding Officer: LtCol Jason L. Morris

1st Battalion, 8th Marines [1st Bn, 8th Mar]
Commanding Officer: LtCol Daniel T. Canfield Jr.

2d Battalion, 8th Marines [2d Bn, 8th Mar]
Commanding Officer: LtCol John D. Harrill III

3d Battalion, 8th Marines [3d Bn, 8th Mar]
Commanding Officer: LtCol Farrell J. Sullivan

3d Battalion, 9th Marines [3d Bn, 9th Mar]
Commanding Officer: LtCol David W. Hudspeth

3d Light Armored Reconnaissance Battalion [3d LAR]
Commanding Officer: LtCol Kenneth R. Kassner

2d Reconnaissance Battalion [2d Recon]
Commanding Officer: LtCol Lawrence K. Hussey

Marine Aviation Combat Element

3d Marine Aircraft Wing (Forward) [3d MAW (Fwd)]
Commanding General: BGen Andrew W. O'Donnell Jr.
Assistant Commander: Col Eric K. Fippinger
Chief of Staff: Col Shaun L. Sadler
Sergeant Major: SgtMaj Anthony A. Spadaro

G-1: Maj Kristin L. McCann
G-2: LtCol George J. David (until January 2011)
Maj Ryan C. Shaffer
G-3: Col Jack P. Monroe
G-4: LtCol James A. Coler
G-6: LtCol Robert K. Maldonado

Marine Air Control Group 38 (Forward) [MACG-38 (Fwd)]
Commanding Officer: LtCol Thomas J. Dodds

Marine Heavy Helicopter Squadron 361 [HMH-361]
Commanding Officer: LtCol Douglas V. Glasgow

Marine Heavy Helicopter Squadron 362 [HMH-362]
Commanding Officer: LtCol Thomas A. Pecina

Marine Heavy Helicopter Squadron 461 [HMH-461]
Commanding Officer: LtCol Scott W. Wadle

Marine Heavy Helicopter Squadron 463 [HMH-463]
Commanding Officer: LtCol Peter S. Gadd

Marine Light Attack Helicopter Squadron 169 [HMLA-169]
Commanding Officer: LtCol Brendan Reilly

Marine Fighter Attack Squadron 122 [VMFA-122]
Commanding Officer: LtCol John A. Bolt

Marine Aerial Refueler Transport Squadron 252 (-) [VMGR-252 (-)]
Commanding Officer: Maj Brent A. Johnson (until January 2011)
Maj Scott M. Koltick

Marine Medium Tiltrotor Squadron 264 [VMM-264]
Commanding Officer: LtCol Jeffrey P. Hogan

Marine Medium Tiltrotor Squadron 266 (Reinforced) [VMM-266 (Rein)]
Commanding Officer: LtCol Romin Dasmalchi

Marine Medium Tiltrotor Squadron 365 [VMM-365]
Commanding Officer: LtCol Craig C. LeFlore

Marine Unmanned Aerial Vehicle Squadron 2 [VMU-2]
Commanding Officer: LtCol Daniel T. Lathrop

Marine Aviation Logistics Squadron 16 [MALS-16]
Commanding Officer: LtCol James C. Carroll III

Marine Wing Support Squadron 373 [MWSS-373]
Commanding Officer: LtCol Thomas M. Fahy

II Marine Expeditionary Force (Forward) [II MEF (Fwd)]

(March–December 2011)

Commanding General: MajGen W. John A. Toolan Jr.
Deputy Commander: BGen Nicholas Welch, UK-A
Chief of Staff: Col Lawrence E. Killmeier Jr.
Sergeant Major: SgtMaj Michael F. Jones

C-1: Col Mary A. Augustitus*
C-2: Col Dave W. Burton
C-3: Col Norman L. Cooling
C-4: Col Jeffrey M. Reagan
C-6: Col David W. McMorries

Command Element

1st Battalion, 23d Marines [1st Bn, 23d Mar]
Commanding Officer: LtCol R. Todd Zink

1st Battalion, 25th Marines [1st Bn, 25th Mar]
Commanding Officer: LtCol Brian O'Leary

2d Intelligence Battalion (Forward) [2d Intel Bn (Fwd)]
Commanding Officer: LtCol Joseph Gross

2d Radio Battalion (Forward) [2d Radio Bn (Fwd)]
Commanding Officer: LtCol Matthew G. Rau

* In the U.S. military, the commanding officer's rank determines whether the staff sections are designated by a G or an S. Organizations commanded by generals have G staffs. Other organizations have S staffs. Two exceptions are joint commands, which contain forces from multiple Services, and are designated with a J, and combined commands, which contain multinational forces and are designated with a C.

8th Communication Battalion (Forward) [8th Comm Bn (Fwd)]
Commanding Officer: LtCol Michael B. Schweighardt

1st Air Naval Gunfire Liaison Company [1st ANGLICO]
Commanding Officer: LtCol Brian T. Bruggeman

2d Air Naval Gunfire Liaison Company [2d ANGLICO]
Commanding Officer: LtCol Leland W. Suttee

Marine Ground Combat Element

2d Marine Division (Forward) [2d MarDiv (Fwd)]
Commanding General: BGen Lewis A. Craparotta
Chief of Staff: Col Andrew H. Smith
Sergeant Major: SgtMaj Joseph D. Shaw

G-1: LtCol Keith F. Moore
G-2: LtCol Robert J. Plevell (until July 2011)
Col Richard E. Anders
G-3: Col Kenneth M. DeTreux
G-4: Col Kyle J. Nickel
G-6: LtCol Dudley R. Griggs

Regimental Combat Team 1 [RCT-1]
Commanding Officer: Col David J. Furness
Executive Officer: LtCol Donald Wright
Sergeant Major: SgtMaj Bryan K. Zickefoose

Regimental Combat Team 5 [RCT-5]
Commanding Officer: Col Roger B. Turner Jr.
Executive Officer: LtCol Peter Wilson

Regimental Combat Team 8 [RCT-8]
Commanding Officer: Col Eric M. Smith
Executive Officer: LtCol Steven G. Luhrsen
Sergeant Major: SgtMaj David L. Jobe

2d Battalion, 1st Marines [2d Bn, 1st Mar]
Commanding Officer: LtCol Matthew S. Reid

3d Battalion, 2d Marines [3d Bn, 2d Mar]
Commanding Officer: LtCol Christopher G. Dixon (until December 2011)
LtCol Charles R. Cassidy

1st Battalion, 3d Marines [1st Bn, 3d Mar]
Commanding Officer: LtCol Sean M. Riordan

2d Battalion, 3d Marines [2d Bn, 3d Mar]
Commanding Officer: LtCol John W. Evans Jr.

3d Battalion, 3d Marines [3d Bn, 3d Mar]
Commanding Officer: LtCol Matthew J. Palma

2d Battalion, 4th Marines [2d Bn, 4th Mar]
Commanding Officer: LtCol William H. Vivian

1st Battalion, 5th Marines [1st Bn, 5th Mar]
Commanding Officer: LtCol Thomas B. Savage

3d Battalion, 5th Marines [3d Bn, 5th Mar]
Commanding Officer: LtCol Jason L. Morris

1st Battalion, 6th Marines [1st Bn, 6th Mar]
Commanding Officer: LtCol George S. Benson

2d Battalion, 6th Marines [2d Bn, 6th Mar]
Commanding Officer: LtCol Matthew T. Morrissey

3d Battalion, 6th Marines [3d Bn, 6th Mar]
Commanding Officer: LtCol Daniel A. Schmitt

3d Battalion, 7th Marines [3d Bn, 7th Mar]
Commanding Officer: LtCol Seth W. B. Folsom

1st Battalion, 8th Marines [1st Bn, 8th Mar]
Commanding Officer: LtCol Daniel T. Canfield

2d Battalion, 8th Marines [2d Bn, 8th Mar]
Commanding Officer: LtCol John D. Harrill III

3d Battalion, 8th Marines [3d Bn, 8th Mar]
Commanding Officer: LtCol Farrell J. Sullivan

1st Battalion, 9th Marines [1st Bn, 9th Mar]
Commanding Officer: LtCol Tyler J. Zagurski

2d Battalion, 9th Marines [2d Bn, 9th Mar]
Commanding Officer: LtCol Michael S. Styskal

3d Battalion, 9th Marines [3d Bn, 9th Mar]
Commanding Officer: LtCol David W. Hudspeth

1st Battalion, 10th Marines [1st Bn, 10th Mar]
Commanding Officer: LtCol Jeffrey C. Smitherman

1st Battalion, 12th Marines [1st Bn, 12th Mar]
Commanding Officer: LtCol Sean S. Charney

2d Combat Engineer Battalion [2d CEB]
Commanding Officer: LtCol Andrew M. Niebel (until April 2011)
LtCol Eric R. Quehl

3d Combat Engineer Battalion [3d CEB]
Commanding Officer: LtCol John P. Sullivan Jr.

1st Light Armored Reconnaissance Battalion [1st LAR]
Commanding Officer: LtCol George C. Schreffler III

3d Light Armored Reconnaissance Battalion [3d LAR]
Commanding Officer: LtCol Kenneth R. Kassner

2d Reconnaissance Battalion [2d Recon]
Commanding Officer: LtCol Lawrence K. Hussey

3d Reconnaissance Battalion [3d Recon]
Commanding Officer: LtCol Travis L. Homiak

Marine Aviation Combat Element

2d Marine Aircraft Wing (Forward) [2d MAW (Fwd)]
Commanding General: BGen Glenn M. Walters

Assistant Commander: Col Russell A. C. Sanborn (until May 2011)
Col Ben Hancock
Chief of Staff: Col Ben Hancock (until May 2011)
Col Mary J. McGregor (until June 2011)
Col James L. Parker
Sergeant Major: SgtMaj Henry A. Prutch II

G-1: Maj Harry L. Gardner
G-2: LtCol Jeffrey S. Stimpson
G-3: Col Jefferson L. DuBinok
G-4: Col Mary J. MacGregor (until May 2011)
LtCol Tyson B. Geisendorff
G-6: Maj Luke L. Fabiunke

Marine Air Control Group 28 (-) [MACG-28 (-)]
Commanding Officer: LtCol Thomas P. Bajus II (until September 2011)
LtCol John R. Siary

Marine Heavy Helicopter Squadron 363 [HMH-363]
Commanding Officer: LtCol Mark S. Revor

Marine Heavy Helicopter Squadron 463 [HMH-463]
Commanding Officer: LtCol Peter S. Gadd

Marine Heavy Helicopter Squadron 464 [HMH-464]
Commanding Officer: LtCol Alison J. Thompson

Marine Light Attack Helicopter Squadron 169 [HMLA-169]
Commanding Officer: LtCol Brendan Reilly

Marine Light Attack Helicopter Squadron 267 [HMLA-267]
Commanding Officer: LtCol Matthew T. Mowery

Marine Light Attack Helicopter Squadron 369 (-) [HMLA-369 (-)]
Commanding Officer: LtCol Ian R. Clark

Marine Attack Squadron 223 [VMA-223]
Commanding Officer: LtCol Thomas D. Gore

Marine Fighter Attack Squadron 122 [VMFA-122]

Commanding Officer: LtCol John A. Bolt

Marine Fighter Attack Squadron 513 [VMFA-513]
Commanding Officer: LtCol Peter N. Lee

Marine Aerial Refueler Transport Squadron 252 (-) [VMGR-252 (-)]
Commanding Officer: Maj Scott M. Koltick (until July 2011)
Maj Walter J. Butler

Marine Medium Tiltrotor Squadron 162 [VMM-162]
Commanding Officer: LtCol Michael S. Ducar

Marine Medium Tiltrotor Squadron 264 [VMM-264]
Commanding Officer: LtCol Jeffrey P. Hogan (until April 2011)
LtCol Brian G. McAvoy

Marine Unmanned Aerial Vehicle Squadron 1 [VMU-1]
Commanding Officer: LtCol John B. Barranco

Marine Unmanned Aerial Vehicle Squadron 2 [VMU-2]
Commanding Officer: LtCol Daniel T. Lathrop (until April 2011)
LtCol Mikel R. Huber

Marine Unmanned Aerial Vehicle Squadron 3 [VMU-3]
Commanding Officer: LtCol Timothy G. Burton

Marine Aviation Logistics Squadron 40 [MALS-40]
Commanding Officer: LtCol Clarence T. Harper III (until July 2011)
LtCol Russell A. Blauw

Marine Wing Support Squadron 272 [MWSS-272]
Commanding Officer: LtCol Dale R. Kruse

Marine Logistics Combat Element

2d Marine Logistics Group (Forward) [2d MLG (Fwd)]
Commanding General: BGen Michael G. Dana
Chief of Staff: Col Mark R. Hollahan (until October 2011)
Col Kenneth D. Enzor
Sergeant Major: SgtMaj William T. Stables

G-1: Maj Michael D. Porter (until August 2011)
Capt Paul W. Steketee (until October 2011)
Maj Beth S. Canepa
G-2: Maj Matthew D. Regner
G-3: Col David W. Maxwell
G-4: Maj Robert J. Livingston Jr.
G-6: Maj Joseph D. Broome (until August 2011)
Maj Donald J. Barnes

Combat Logistics Battalion 1 [CLB-1]
Commanding Officer: LtCol Brian H. Kane (until November 2011)
Maj Jason A. Gaddy (from November 2011)
LtCol Michael W. Stehle

Combat Logistics Battalion 3 [CLB-3]
Commanding Officer: LtCol Henry W. Lutz III

Combat Logistics Battalion 6 [CLB-6]
Commanding Officer: LtCol Ralph J. Rizzo Jr.

Combat Logistics Battalion 7 [CLB-7]
Commanding Officer: LtCol Craig P. Barnett

Combat Logistics Battalion 8 [CLB-8]
Commanding Officer: LtCol Michael E. McWilliams

7th Engineer Support Battalion [7th Eng Spt Bn]
Commanding Officer: LtCol Daniel H. Dubbs

8th Engineer Support Battalion [8th Eng Spt Bn]
Commanding Officer: LtCol Christopher G. Downs

9th Engineer Support Battalion [9th Eng Spt Bn]
Commanding Officer: LtCol Scott A. Baldwin

Marine Air-Ground Task Force Support Battalion [MAGTF Spt Bn]
Commanding Officer: LtCol Kevin R. Scott (until September 2011)
LtCol David S. Gibbs

2012

International Security Assistance Force, United States Forces-Afghanistan [ISAF USF-A]

9th Air and Space Expeditionary Task Force-Afghanistan [9th AETF-A]

Marine Tactical Electronic Warfare Squadron 1 [VMAQ-1]
Commanding Officer: LtCol Chandler P. Seagraves

Marine Tactical Electronic Warfare Squadron 2 [VMAQ-2]
Commanding Officer: LtCol Adam L. Musoff

International Security Assistance Force, Regional Command-Southwest [ISAF RC-SW]

II Marine Expeditionary Force (Forward) [II MEF Fwd)]

(January–March 2012)

Commanding General: MajGen John A. Toolan
Deputy Commander: BGen Nicholas Welch, UK-A
Chief of Staff: Col Lawrence E. Killmeier Jr.
Sergeant Major: SgtMaj Michael F. Jones

C-1: Col Mary A. Augustitus
C-2: Col Dave W. Burton
C-3: Col Norman L. Cooling
C-4: Col Jeffrey M. Reagan
C-6: Col David W. McMorries

Command Element

1st Battalion, 25th Marines [1st Bn, 25th Mar]
Commanding Officer: LtCol Brian O'Leary

2d Intelligence Battalion [2d Intel Bn]
Commanding Officer: LtCol Joseph Gross

2d Radio Battalion [2d Radio Bn]
Commanding Officer: LtCol Matthew G. Rau

8th Communication Battalion [8th Comm Bn]
Commanding Officer: LtCol Michael B. Schweighardt

5th Air Naval Gunfire Liaison Company [5th ANGLICO]
Commanding Officer: LtCol Kendall A. Martinez

Marine Ground Combat Element

2d Marine Division (Forward) [2d MarDiv (Fwd)]
Commanding General: BGen Lewis A. Craparotta
Chief of Staff: Col Andrew H. Smith
Sergeant Major: SgtMaj Joseph D. Shaw

G-1: LtCol Keith F. Moore
G-2: Col Richard E. Anders
G-3: Col Kenneth M. DeTreux
G-4: Col Kyle J. Nickel
G-6: LtCol Dudley R. Griggs

Regimental Combat Team 5 [RCT-5]
Commanding Officer: Col Roger B. Turner Jr.
Executive Officer: LtCol Peter C. Wilson
Sergeant Major: SgtMaj Ernest K. Hoopii

Regimental Combat Team 6 [RCT-6]
Commanding Officer: Col John R. Shafer
Executive Officer: LtCol Bradley C. Weston
Sergeant Major: SgtMaj Jamie A. Deets

Regimental Combat Team 8 [RCT-8]
Commanding Officer: Col Eric M. Smith
Executive Officer: LtCol Steven G. Luhrsen
Sergeant Major: SgtMaj David L. Jobe

3d Battalion, 3d Marines [3d Bn, 3d Mar]
Commanding Officer: LtCol Matthew Palma

2d Battalion, 4th Marines [2d Bn, 4th Mar]
Commanding Officer: LtCol William H. Vivian

1st Battalion, 6th Marines [1st Bn, 6th Mar]
Commanding Officer: LtCol George S. Benson

2d Battalion, 6th Marines [2d Bn, 6th Mar]
Commanding Officer: LtCol Matthew T. Morrissey

3d Battalion, 7th Marines [3d Bn, 7th Mar]
Commanding Officer: LtCol Seth W. B. Folsom

1st Battalion, 8th Marines [1st Bn, 8th Mar]
Commanding Officer: LtCol Kevin C. Trimble

2d Battalion, 9th Marines [2d Bn, 9th Mar]
Commanding Officer: LtCol Michael S. Styskal

2d Battalion, 11th Marines [2d Bn, 11th Mar]
Commanding Officer: LtCol Joseph T. Allena Jr.

3d Combat Engineer Battalion [3d CEB]
Commanding Officer: LtCol John P. Sullivan Jr.

1st Light Armored Reconnaissance Battalion [1st LAR]
Commanding Officer: LtCol George C. Schreffler III

1st Reconnaissance Battalion [1st Recon Bn]
Commanding Officer: LtCol Brian L. Gilman

Marine Aviation Combat Element

2d Marine Aircraft Wing (Forward) [2d MAW (Fwd)]
Commanding General: BGen Glenn M. Walters
Assistant Commander: Col Ben Hancock
Chief of Staff: Col James L. Parker
Sergeant Major: SgtMaj Henry A. Prutch II

G-1: Maj Harry L. Gardner
G-2: LtCol Jeffrey S. Stimpson
G-3: Col Jefferson L. DuBinok
G-4: LtCol Tyson B. Geisendorff
G-6: Maj Luke L. Fabiunke

Marine Air Control Group 28 (-) [MACG-28 (-)]
Commanding Officer: LtCol John R. Siary

Marine Heavy Helicopter Squadron 363 [HMH-363]
Commanding Officer: LtCol Mark S. Revor

Marine Heavy Helicopter Squadron 464 [HMH-464]
Commanding Officer: LtCol Alison J. Thompson

Marine Light Attack Helicopter Squadron 369 [HMLA-369]
Commanding Officer: LtCol Ian R. Clark

Marine Attack Squadron 223 [VMA-223]
Commanding Officer: LtCol Thomas D. Gore

Marine Medium Tiltrotor Squadron 365 [VMM-365]
Commanding Officer: LtCol Craig C. LeFlore

Marine Unmanned Aerial Vehicle Squadron 1 [VMU-1]
Commanding Officer: LtCol John B. Barranco

Marine Aviation Logistics Squadron 40 (Forward) [MALS-40 (Fwd)]
Commanding Officer: LtCol Russell A. Blauw

Marine Wing Support Squadron 371 [MWSS-371]
Commanding Officer: LtCol Michael B. Prosser

Marine Logistics Combat Element

1st Marine Logistics Group (Forward) [1st MLG (Fwd)]
Commanding General: BGen John J. Broadmeadow
Chief of Staff: Col Stephen D. Sklenka
Sergeant Major: SgtMaj Antonio N. Vizcarrondo Jr.

G-1: Maj Beth S. Canepa (until February 2012)
Capt Alexandra L. Cole
G-2: Capt Keith S. Crim
G-3: Col William B. Pitman
G-4: Maj Rugithi D. Meelarp
G-6: Maj Donald J. Barnes (until February 2012)
Maj Donald A. Hudson

Combat Logistics Battalion 1 [CLB-1]
Commanding Officer: LtCol Michael W. Stehle

Combat Logistics Battalion 4 [CLB-4]
Commanding Officer: LtCol Adam L. Chalkey

9th Engineer Support Battalion [9th Eng Spt Bn]
Commanding Officer: LtCol Scott A. Baldwin

MAGTF Support Battalion [MAGTF Spt Bn]
Commanding Officer: LtCol David S. Gibbs

I Marine Expeditionary Force (Forward) [I MEF (Fwd)]

(March–December 2012)

Commanding General: MajGen Charles M. Gurganus
Deputy Commander: BGen Stuart R. Skeates, UK-A
Chief of Staff: Col Timothy W. Fitzgerald
Sergeant Major: SgtMaj Harrison L. Tanksley

C-1: Col James P. Rethwisch
C-2: Col Timothy J. Oliver
C-3: Col Jeffrey D. Tuggle
C-4: Col Jeffrey Q. Hooks
C-6: LtCol Roger D. Standfield

Command Element

1st Intelligence Battalion [1st Intel Bn]
Commanding Officer: LtCol Dawn R. Alonso

1st Radio Battalion [1st Radio Bn]
Commanding Officer: LtCol Ahmed T. Williamson

9th Communication Battalion [9th Comm Bn]
Commanding Officer: Maj Joseph T. Dellos

Marine Ground Combat Element

1st Marine Division (Forward) [1st MarDiv (Fwd)]
Commanding General: MajGen David H. Berger

Chief of Staff: Col Ronald A. Gridley
Sergeant Major: SgtMaj Terry L. Jones

G-1: LtCol Jerry R. Morgan
G-2: LtCol Michael W. Taylor
G-3: Col Jeffrey T. Conner
G-4: Col Thomas B. Eipp
G-6: Col Joseph A. Cabell

Regimental Combat Team 5 [RCT-5]
Commanding Officer: Col Roger B. Turner Jr.
Executive Officer: LtCol Minter B. Ralston IV
Sergeant Major: SgtMaj Alberto Ruiz

Regimental Combat Team 6 [RCT-6]
Commanding Officer: Col John R. Shafer
Executive Officer: LtCol Bradley C. Weston
Sergeant Major: SgtMaj Jamie A. Deets

Regimental Combat Team 7 [RCT-7]
Commanding Officer: Col Austin E. Renforth
Executive Officer: LtCol Brian W. Neil
Sergeant Major: SgtMaj Scott A. Samuels

1st Battalion, 1st Marines [1st Bn, 1st Mar]
Commanding Officer: LtCol Michael J. Targos

3d Battalion, 3d Marines [3d Bn, 3d Mar]
Commanding Officer: LtCol Matthew J. Palma

2d Battalion, 5th Marines [2d Bn, 5th Mar]
Commanding Officer: LtCol Jason S. D. Perry (until December 2012)
LtCol Timothy M. Bairstow

2d Battalion, 6th Marines [2d Bn, 6th Mar]
Commanding Officer: LtCol Matthew T. Morrissey

1st Battalion, 7th Marines [1st Bn, 7th Mar]
Commanding Officer: LtCol David P. Bradney

2d Battalion, 7th Marines [2d Bn, 7th Mar]
Commanding Officer: LtCol Donald J. Tomich

3d Battalion, 7th Marines [3d Bn, 7th Mar]
Commanding Officer: LtCol Seth W. B. Folsom

1st Battalion, 8th Marines [1st Bn, 8th Mar]
Commanding Officer: LtCol Kevin C. Trimble

3d Battalion, 8th Marines [3d Bn, 8th Mar]
Commanding Officer: LtCol Edward J. Healey

2d Battalion, 9th Marines [2d Bn, 9th Mar]
Commanding Officer: LtCol Michael S. Styskal

3d Battalion, 9th Marines [3d Bn, 9th Mar]
Commanding Officer: LtCol Carl E. Cooper Jr.

2d Battalion, 10th Marines [2d Bn, 10th Mar]
Commanding Officer: LtCol Phillip M. Bragg

2d Battalion, 11th Marines [2d Bn, 11th Mar]
Commanding Officer: LtCol Joseph T. Allena Jr.

1st Combat Engineer Battalion [1st CEB]
Commanding Officer: LtCol Andrew M. Niebel

3d Combat Engineer Battalion [3d CEB]
Commanding Officer: LtCol John P. Sullivan Jr.

1st Light Armored Reconnaissance Battalion [1st LAR]
Commanding Officer: LtCol George C. Schreffler III

3d Light Armored Reconnaissance Battalion [3d LAR]
Commanding Officer: LtCol Mark H. Clingan

1st Reconnaissance Battalion [1st Recon Bn]
Commanding Officer: LtCol Brian L. Gilman

Marine Aviation Combat Element

3d Marine Aircraft Wing (Forward) [3d MAW (Fwd)]
Commanding General: BGen Gregg A. Sturdevant
Chief of Staff: Col Jeffrey P. Davis (until May 2012)
Col Richard T. Bew
Sergeant Major: SgtMaj Rodolfo J. Arrieta

G-1: Maj Karen B. Brockmeier (until August 2012)
Capt Steven D. Harvey
G-2: Maj Ryan C. Shaffer (until August 2012)
Capt Karen R. Reyes
G-3: Col Christopher A. Keane
G-4: Col Wayne R. Steele (until June 2012)
Maj Peter E. Dahl (until August 2012)
LtCol Jose M. Lopez
G-6: LtCol Robert K. Maldonado

Marine Air Control Group 38 (-) (Forward) [MACG-38 (-) (Fwd)]
Commanding Officer: LtCol Scott A. Gondek (until September 2012)
LtCol John C. Lewis

Marine Heavy Helicopter Squadron 361 [HMH-361]
Commanding Officer: LtCol Jay M. Holtermann

Marine Heavy Helicopter Squadron 362 [HMH-362]
Commanding Officer: LtCol Christopher H. Oliver

Marine Heavy Helicopter Squadron 363 [HMH-363]
Commanding Officer: LtCol Mark S. Revor

Marine Heavy Helicopter Squadron 466 [HMH-466]
Commanding Officer: LtCol Joseph K. Decapite

Marine Light Attack Helicopter Squadron 169 [HMLA-169]
Commanding Officer: LtCol Garret R. Hoffman

Marine Light Attack Helicopter Squadron 369 [HMLA-369]
Commanding Officer: LtCol Ian R. Clark

Marine Light Attack Helicopter Squadron 469 [HMLA-469]
Commanding Officer: LtCol Stephen J. Lightfoot

Marine Attack Squadron 211 [VMA-211]
Commanding Officer: LtCol Christopher K. Raible (until September 2012)
Maj John F. Havener (until September 2012)
LtCol Troy M. Pehrson

Marine Attack Squadron 223 [VMA-223]
Commanding Officer: LtCol Thomas D. Gore

Marine Attack Squadron 231 [VMA-231]
Commanding Officer: LtCol Shawn R. Hermley

Marine Aerial Refueler Transport Squadron 352 (-) [VMGR-352 (-)]
Commanding Officer: Maj Daniel R. Campbell (until July 2012)
Maj Drew R. Hess

Marine Medium Tiltrotor Squadron 161 [VMM-161]
Commanding Officer: LtCol Eric M. Gillard

Marine Medium Tiltrotor Squadron 365 [VMM-365]
Commanding Officer: LtCol Craig C. LeFlore

Marine Unmanned Aerial Vehicle Squadron 1 [VMU-1]
Commanding Officer: LtCol John B. Barranco

Marine Unmanned Aerial Vehicle Squadron 2 [VMU-2]
Commanding Officer: LtCol Mikel R. Huber (until September 2012)
Maj Jeremy G. Hall

Marine Unmanned Aerial Vehicle Squadron 3 [VMU-3]
Commanding Officer: Maj Daniel Lindblom

Marine Aviation Logistics Squadron 16 (Forward) [MALS-16 (Fwd)]
Commanding Officer: LtCol Raymond E. Barnett (until August 2012)
LtCol Jason S. Guello

Marine Wing Support Squadron 273 [MWSS-273]
Commanding Officer: LtCol William S. Kohmuench

Marine Wing Support Squadron 373 [MWSS-373]
Commanding Officer: LtCol Seth L. Ocloo
Maj Gregory K. Butler

Marine Logistics Combat Element

1st Marine Logistics Group (Forward) [1st MLG (Fwd)]
Commanding General: BGen John J. Broadmeadow
Chief of Staff: Col Stephen D. Sklenka
Sergeant Major: SgtMaj Antonio N. Vizcarrondo Jr.

G-1: Capt Alexandra L. Cole (until Jul 2012)
Maj Brian J. Brauer
G-2: Capt Keith S. Crim
G-3: Col William B. Pitman
G-4: Maj Rugithi D. Meelarp
G-6: Maj Donald A. Hudson

Combat Logistics Battalion 1 [CLB-1]
Commanding Officer: LtCol Michael W. Stehle

Combat Logistics Battalion 2 [CLB-2]
Commanding Officer: LtCol Denise M. Mull

Combat Logistics Battalion 4 [CLB-4]
Commanding Officer: LtCol Adam L. Chalkey (until November 2012)
LtCol Travis T. Gaines

Combat Logistics Battalion 5 [CLB-5]
Commanding Officer: LtCol Robert T. Meade

1st Maintenance Battalion (-) (Reinforced) [1st Mnt Bn (-)(Rein)]
Commanding Officer: LtCol Christian D. Richardson

8th Engineer Support Battalion [8th Eng Spt Bn]
Commanding Officer: LtCol Ferdinand F. Llantero

9th Engineer Support Battalion [9th Eng Spt Bn]
Commanding Officer: LtCol Scott A. Baldwin

2013

9th Air and Space Expeditionary Task Force-Afghanistan [9th AETF-A]

Marine Tactical Electronic Warfare Squadron 4 [VMAQ-4]
Commanding Officer: LtCol Tommy D. Cornstubble

International Security Assistance Force, Regional Command-Southwest [ISAF RC-SW]

I Marine Expeditionary Force (Forward) [I MEF (Fwd)]

(January–February 2013)

Commanding General: MajGen Charles M. Gurganus
Deputy Commander: BGen Stewart R. Skeates, UK-A
Chief of Staff: Col Timothy W. Fitzgerald
Sergeant Major: SgtMaj Harrison L. Tanksley

C-1: Col James P. Rethwisch
C-2: Col Timothy J. Oliver
C-3: Col Jeffrey D. Tuggle
C-4: Col Jeffrey Q. Hooks
C-6: LtCol Roger D. Standfield

Command Element

1st Intelligence Battalion [1st Intel Bn]
Commanding Officer: LtCol Dawn Alonso

1st Radio Battalion [1st Radio Bn]
Commanding Officer: LtCol Ahmed Williamson

9th Communication Battalion [9th Comm Bn]
Commanding Officer: Maj Joseph Dellos

Marine Ground Combat Element

Regimental Combat Team 7 [RCT-7]
Commanding Officer: Col Austin E. Renforth
Executive Officer: LtCol Brian W. Neil
Sergeant Major: SgtMaj Scott A. Samuels

2d Battalion, 7th Marines [2d Bn, 7th Mar]
Commanding Officer: LtCol Donald J. Tomich

3d Battalion, 9th Marines [3d Bn, 9th Mar]
Commanding Officer: LtCol Carl E. Cooper Jr.

2d Battalion, 10th Marines [2d Bn, 10th Mar]
Commanding Officer: LtCol Phillip M. Bragg

2d Combat Engineer Battalion [2d CEB]
Commanding Officer: LtCol John C. Osborne Jr.

Marine Aviation Combat Element

3d Marine Aircraft Wing (Forward) [3d MAW (Fwd)]
Commanding General: BGen Gregg A. Sturdevant
Chief of Staff: Col Richard T. Bew
Sergeant Major: SgtMaj Rodolfo J. Arrieta

G-1: Capt Steven D. Harvey
G-2: Capt Karen R. Reyes
G-3: Col Christopher A. Keane
G-4: LtCol Jose M. Lopez
G-6: LtCol Robert K. Maldonado

Marine Air Control Group 28 (Forward) [MACG-28 (Fwd)]
Commanding Officer: LtCol Darry W. Grossnickle

Marine Heavy Helicopter Squadron 361 [HMH-361]
Commanding Officer: LtCol Jay M. Holtermann

Marine Heavy Helicopter Squadron 461 [HMH-461]
Commanding Officer: LtCol Nicholas A. Morris

Marine Light Attack Helicopter Squadron 169 [HMLA-169]
Commanding Officer: LtCol Garrett R. Hoffman

Marine Attack Squadron 211 [VMA-211]
Commanding Officer: LtCol Troy M. Pehrson

Marine Attack Squadron 231 [VMA-231]
Commanding Officer: LtCol S. R. Hermley (until June 2013)
LtCol M. P. Quinto (from June 2013)

Marine Aerial Refueler Transport Squadron 252 (-) [VMGR-252 (-)]
Commanding Officer: Maj James C. Paxton III

Marine Medium Tiltrotor Squadron 161 [VMM-161]
Commanding Officer: LtCol Douglas C. Sanders

Marine Unmanned Aerial Vehicle Squadron 3 (Forward) [VMU-3 (Fwd)]
Commanding Officer: Maj Daniel E. Lindblom

Marine Aviation Logistics Squadron 40 [MALS-40]
Commanding Officer: LtCol Ryan G. Goulette

Marine Wing Support Squadron 373 (Forward) [MWSS-373 (Fwd)]
Commanding Officer: Maj Gregory K. Butcher

Marine Logistics Combat Element

Combat Logistics Battalion 2 [CLB-2]
Commanding Officer: LtCol Denise M. Mull

II Marine Expeditionary Force (Forward) [II MEF (Fwd)]

(February–December 2013)

Commanding General: MajGen W. Lee Miller Jr.
Deputy Commander: BGen Paul A. E. Nanson, UK-A
Chief of Staff: Col James Bright (until June 2013)
Col Barry J. Fitzpatrick Jr.
Sergeant Major: SgtMaj Paul A. Berry

C-1: LtCol Maureen Murphy
C-2: LtCol Vernon Williams (until July 2013)
LtCol Andrew J. Weis
C-3: Col James E. Donnellan
C-4: Col Michael H. Oppenheim
C-6: LtCol Raul L. Salcido

Command Element

2d Law Enforcement Battalion [2d LEB]
Commanding Officer: LtCol Eric W. Young

Marine Ground Combat Element

2d Marine Regiment [2d Mar]
Commanding Officer: Col Daniel M. Sullivan
Executive Officer: LtCol David M. Fallon
Sergeant Major: SgtMaj Paul T. Archie (until December 2013)
SgtMaj James M. Boutin

Regimental Combat Team 7 [RCT-7]
Commanding Officer: Col Austin E. Renforth
Executive Officer: LtCol Brian W. Neil (until May 2013)
LtCol Christian M. Rankin
Sergeant Major: SgtMaj Scott A. Samuels

3d Battalion, 4th Marines [3d Bn, 4th Mar]
Commanding Officer: LtCol Jeffrey R. Kenney

2d Battalion, 7th Marines [2d Bn, 7th Mar]
Commanding Officer: LtCol Donald J. Tomich

3d Battalion, 7th Marines [3d Bn, 7th Mar]
Commanding Officer: LtCol Daniel J. Wittnam

2d Battalion, 8th Marines [2d Bn, 8th Mar]
Commanding Officer: LtCol Kevin E. Clark

3d Battalion, 9th Marines [3d Bn, 9th Mar]
Commanding Officer: LtCol Carl E. Cooper Jr.

1st Combat Engineer Battalion (-) (Reinforced) [1st CEB (-) (Rein)]
Commanding Officer: LtCol Andrew R. Winthrop

2d Combat Engineer Battalion [2d CEB]
Commanding Officer: LtCol John C. Osborne Jr.

3d Combat Engineer Battalion (Forward) [3d CEB (Fwd)]
Commanding Officer: LtCol Brian M. Dwyer

Marine Aviation Combat Element

2d Marine Aircraft Wing (Forward) [2d MAW (Fwd)]
Commanding General: BGen Gary L. Thomas (until December 2013)
Commanding Officer: Col Scott S. Jensen (from December 2013)
Assistant Commander: Col Kevin M. Iiams (until April 2013)
Col John C. Vara (until May 2013)
Col Scott S. Jensen (until December 2013)
LtCol Thomas P. Bajus II
Chief of Staff: LtCol Thomas P. Bajus II (from November 2013)
Sergeant Major: SgtMaj Lisa K. Nilsson

G-1: Capt Kerrissa A. Sterns
G-2: LtCol Fred W. Bista (until July 2013)
Capt Tristan J. Young
G-3: Col John C. Vara (until July 2013)
LtCol Mark E. Van Sike
G-4: LtCol Paul R. Milne
G-6: Maj Randall G. Turner (until July 2013)
Maj Miguel A. Guerra

Marine Air Control Group 28 (Forward) [MACG-28 (Fwd)]
Commanding Officer: LtCol Darry W. Grossnickle (until July 2013)
LtCol Charles E. Smith

Marine Heavy Helicopter Squadron 361 [HMH-361]
Commanding Officer: LtCol Jay M. Holtermann

Marine Heavy Helicopter Squadron 461 [HMH-461]
Commanding Officer: LtCol Nicholas A. Morris

Marine Heavy Helicopter Squadron 462 [HMH-462]
Commanding Officer: LtCol Jacob M. Matt

Marine Light Attack Helicopter Squadron 167 [HMLA-167]
Commanding Officer: LtCol Shawn J. Hughes (until August 2013)
LtCol William J. Bartolema (until December 2013)
LtCol Robert B. Finneran

Marine Light Attack Helicopter Squadron 169 [HMLA-169]
Commanding Officer: LtCol Garrett R. Hoffman

Marine Light Attack Helicopter Squadron 369 [HMLA-369]
Commanding Officer: LtCol Tres C. Smith

Marine Attack Squadron 231 [VMA-231]
Commanding Officer: LtCol Shawn R. Hermley (until June 2013)
LtCol Michael P. Quinto

Marine Attack Squadron 311 [VMA-311]
Commanding Officer: LtCol Scott C. Mitchell

Marine Aerial Refueler Transport Squadron 252 (-) [VMGR-252 (-)]
Commanding Officer: Maj James C. Paxton III (until July 2013)
Maj Ryan C. Pope

Marine Medium Tiltrotor Squadron 161 [VMM-161]
Commanding Officer: LtCol Douglas C. Sanders

Marine Medium Tiltrotor Squadron 165 [VMM-165]
Commanding Officer: LtCol Kurt J. Schiller

Marine Medium Tiltrotor Squadron 264 [VMM-264]
Commanding Officer: LtCol Douglas C. Sanders

Marine Unmanned Aerial Vehicle Squadron 2 (Forward) [VMU-2 (Fwd)]
Commanding Officer: Maj Philip M. Raymond

Marine Unmanned Aerial Vehicle Squadron 3 (Forward) [VMU-3 (Fwd)]
Commanding Officer: Maj Daniel E. Lindblom

Marine Aviation Logistics Squadron 40 [MALS-40]
Commanding Officer: LtCol Ryan G. Goulette (until July 2013)
LtCol William C. Gray

Marine Wing Support Squadron 271 (-) (Forward) [MWSS-271 (-)(Fwd)]
Commanding Officer: LtCol Erik L. Aubel

Marine Wing Support Squadron 372 (Forward) [MWSS 372 (Fwd)]
Commanding Officer: LtCol Maura M. Hennigan

Marine Logistics Combat Element

Combat Logistics Regiment 2 [CLR-2]

Commanding Officer: Col Dwayne A. Whiteside

Executive Officer:	LtCol James B. Stone IV (until March 2013)
	LtCol Christopher D. Hrudka (until July 2013)
	LtCol Brian W. Mullery
Sergeant Major:	SgtMaj Lannette N. Wright (until July 2013)
	SgtMaj Roger F. Griffith

Combat Logistics Battalion 2 [CLB-2]

Commanding Officer:	LtCol Denise M. Mull (until June 2013)
	LtCol William C. Stophel

2014

9th Air and Space Expeditionary Task Force-Afghanistan [9th AETF-A]

Marine Tactical Electronic Warfare Squadron 4 [VMAQ-4]

Commanding Officer: LtCol Tommy D. Cornstubble

International Security Assistance Force, Regional Command-Southwest [ISAF RC-SW]

II Marine Expeditionary Force (Forward) [II MEF (Fwd)]

(January–February 2014)

Commanding General: MajGen Walter Lee Miller Jr.

Deputy Commander: BGen Paul A. E. Nanson, UK-A

Chief of Staff: Col Barry J. Fitzpatrick

Sergeant Major: SgtMaj Paul A. Berry

C-1:	LtCol Maureen Murphy
C-2:	LtCol Andrew J. Weis
C-3:	Col James E. Donnellan
C-4:	Col Michael H. Oppenheim
C-6:	LtCol Raul L. Salcido

Command Element

2d Law Enforcement Battalion [2d LEB]

Commanding Officer: LtCol Eric W. Young

Marine Ground Combat Element

2d Marine Regiment [2d Mar]
Commanding Officer: Col Daniel M. Sullivan
Executive Officer: LtCol David M. Fallon
Sergeant Major: SgtMaj James M. Boutin

3d Battalion, 7th Marines [3d Bn, 7th Mar]
Commanding Officer: LtCol Daniel J. Wittnam

1st Battalion, 9th Marines [1st Bn, 9th Mar]
Commanding Officer: LtCol Corey M. Collier

1st Combat Engineer Battalion (-) (Reinforced) [1st CEB (-)(Rein)]
Commanding Officer: LtCol Andrew R. Winthrop

Marine Aviation Combat Element

2d Marine Aircraft Wing (Forward) [2d MAW (Fwd)]
Commanding Officer: Col Scott S. Jensen
Assistant Commander: LtCol Thomas P. Bajus II
Sergeant Major: SgtMaj Lisa K. Nilsson

G-1: Capt Kerrissa A. Sterns
G-2: Capt Tristan J. Young
G-3: LtCol Mark E. Van Sike
G-4: LtCol Paul R. Milne
G-6: Maj Miguel A. Guerra

Marine Air Control Group 38 (-) [MACG-38 (-)]
Commanding Officer: LtCol Kevin R. Korpinen

Marine Heavy Helicopter Squadron 462 [HMH-462]
Commanding Officer: LtCol Jacob M. Matt

Marine Light Attack Helicopter Squadron 369 [HMLA-369]
Commanding Officer: LtCol Tres C. Smith

Marine Attack Squadron 311 [VMA-311]
Commanding Officer: LtCol Scott C. Mitchell

Marine Aerial Refueler Transport Squadron 352 (-) [VMGR-352 (-)]
Commanding Officer: Maj Ryan C. Pope (until January 2014)
Maj Casey D. Shea

Marine Medium Tiltrotor Squadron 165 [VMM-165]
Commanding Officer: LtCol Kurt J. Schiller

Marine Unmanned Aerial Vehicle Squadron 2 (-) [VMU-2 (-)]
Commanding Officer: Maj Philip M. Raymond

Marine Air Logistics Squadron 70 [MALS-70]
Commanding Officer: Maj Luke T. Watson

Marine Wing Support Squadron 372 (Forward) [MWSS-372 (Fwd)]
Commanding Officer: LtCol Maura M. Hennigan

Marine Logistics Combat Element

Combat Logistics Battalion 6 [CLB-6]
Commanding Officer: LtCol Brian W. Mullery

Combat Logistics Battalion 7 [CLB-7]
Commanding Officer: LtCol Sidney R. Welch

Marine Expeditionary Brigade-Afghanistan [MEB-A]

(March–November 2014)

Commanding General: BGen Daniel D. Yoo
Chief of Staff: Col Bruce W. Barnhill
Sergeant Major: SgtMaj Douglas E. Berry Jr.

C-1: LtCol Heather J. Cotoia (until April 2014)
LtCol Zachary T. Schmidt
C-2: LtCol Larry M. Jenkins
C-3: Col Anthony M. Henderson
C-4: Col Douglas R. Patterson
C-6: LtCol Kelly K. Hastings

Command Element

1st Radio Battalion (Forward) [1st Radio Bn (Fwd)]

Commanding Officer: Capt David M. Miller (until May 2014)
1stLt Paul J. Kapavik

9th Communication Battalion (Forward) [9th Comm Bn (Fwd)]
Commanding Officer: Maj Tara J. Kipfer (until June 2014)
Capt Matthew Knopp

1st Air Naval Gunfire Liaison Company [1st ANGLICO]
Commanding Officer: LtCol Brian E. Russell

Marine Ground Combat Element

1st Marine Regiment [1st Mar]
Commanding Officer: Col P. B. Baumgarten
Executive Officer: LtCol P. N. Ash (until June 2014)
LtCol R. M. Hoyle
Sergeant Major: SgtMaj R. Robinson

1st Battalion, 2d Marines [1st Bn, 2d Mar]
Commanding Officer: LtCol Bradley C. Weston

3d Battalion, 7th Marines [3d Bn, 7th Mar]
Commanding Officer: LtCol Daniel J. Wittnam

1st Battalion, 9th Marines [1st Bn, 9th Mar]
Commanding Officer: LtCol Corey M. Collier

1st Combat Engineer Battalion (-) (Reinforced) [1st CEB (-)(Rein)]
Commanding Officer: LtCol Andrew R. Winthrop

2d Combat Engineer Battalion [2d CEB]
Commanding Officer: LtCol John C. Osborne Jr.

Marine Aviation Combat Element

Marine Aircraft Group 26 [MAG-26]
Commanding Officer: Col Patrick A. Gramuglia
Executive Officer: LtCol Michael K. Van Nest
Sergeant Major: SgtMaj Mark F. Gonzales

Marine Air Control Group 38 (-) [MACG-38 (-)]
Commanding Officer: LtCol Kevin R. Korpinen

Marine Heavy Helicopter Squadron 462 [HMH-462]
Commanding Officer: LtCol Jacob M. Matt

Marine Heavy Helicopter Squadron 466 [HMH-466]
Commanding Officer: LtCol Timothy A. Sheyda

Marine Light Attack Helicopter Squadron 369 [HMLA-369]
Commanding Officer: LtCol Tres C. Smith

Marine Light Attack Helicopter Squadron 467 [HMLA-467]
Commanding Officer: LtCol Matthew R. Sale

Marine Attack Squadron 311 [VMA-311]
Commanding Officer: LtCol Scott C. Mitchell

Marine Aerial Refueler Transport Squadron 352 (-) [VMGR-352 (-)]
Commanding Officer: Maj Casey D. Shea

Marine Medium Tiltrotor Squadron 261 [VMM-261]
Commanding Officer: LtCol Kelvin W. Gallman

Marine Unmanned Aerial Vehicle Squadron 1 (-) [VMU-1 (-)]
Commanding Officer: Maj Peter Y. Ban

Marine Unmanned Aerial Vehicle Squadron 2 (-) [VMU-2 (-)]
Commanding Officer: Maj Philip M. Raymond

Marine Air Logistics Squadron 70 [MALS-70]
Commanding Officer: Maj Luke T. Watson

Marine Wing Support Squadron 274 (Forward) [MWSS-274 (Fwd)]
Commanding Officer: LtCol Sean C. Killeen

Marine Wing Support Squadron 372 (Forward) [MWSS-372 (Fwd)]
Commanding Officer: LtCol Maura M. Hennigan

Marine Logistics Combat Element

Combat Logistics Battalion 7 [CLB-7]
Commanding Officer: LtCol Sidney R. Welch

Appendix B:
Unit List

U.S. Marines in Operation Enduring Freedom 2010

International Security Assistance Force, U.S. Forces-Afghanistan [ISAF USF-A]

Special Operations Task Force-West

- 1st Marine Special Operations Battalion [1st MSOB]
  - Marine Special Operations Task Force 81 HQ (14 October 2009–31 May 2010)
  - Marine Special Operations Company 811 (25 July 2009–28 Feb 2010)
  - Marine Special Operations Company 812 (6 January 2010–31 July 2010)
  - Marine Special Operations Company 813 (26 August 2010–1 April 2011)
- 2d Marine Special Operations Battalion [2d MSOB]
  - Marine Special Operations Task Force-West HQ (31 May 2010–1 February 2011)
  - Marine Special Operations Company 821 (18 November 2010–1 August 2010)
  - Marine Special Operations Company 822 (14 October 2009–3 June 2010)
  - Marine Special Operations Company 823 (30 April 2010–1 February 2011)

International Security Assistance Force, Regional Command-Southwest [ISAF RC-SW]

2d Marine Expeditionary Brigade [2d MEB]

Command Element

- 2d MEB Headquarters Group
  - 9th Communications Battalion [9th Comm Bn]
  - 1st Battalion, 5th Marines [1st Bn, 5th Mar]
  - 1st Air Naval Gunfire Liaison Company [1st ANGLICO]
  - 4th Civil Affairs Group [4th CAG]

Marine Ground Combat Element

- Regimental Combat Team 2 [RCT-2]
  - 1st Battalion, 2d Marines [1st Bn, 2d Mar]
- Regimental Combat Team 7 [RCT-7]
  - 2d Battalion, 2d Marines [2d Bn, 2d Mar]
  - 1st Battalion, 3d Marines [1st Bn, 3d Mar]

3d Battalion, 4th Marines [3d Bn, 4th Mar]
1st Battalion, 6th Marines [1st Bn, 6th Mar]
3d Battalion, 6th Marines [3d Bn, 6th Mar]
3d Battalion, 10th Marines [3d Bn, 10th Mar]
5th Battalion, 11th Marines [5th Bn, 11th Mar]
4th Light Armored Reconnaissance Battalion [4th LAR]
2d Combat Engineer Battalion [2d CEB]

Marine Aviation Combat Element

Marine Aircraft Group 40 [MAG-40]
Marine Light Attack Helicopter Squadron 367 [HMLA-367]
Marine Aviation Logistics Squadron 40 [MALS-40]
Marine Aviation Logistics Squadron 16 [MALS-16]
Marine Aviation Logistics Squadron 24 [MALS-24]
Marine Aviation Logistics Squadron 26 [MALS-26]
Marine Aviation Logistics Squadron 29 [MALS-29]
Marine Aviation Logistics Squadron 39 [MALS-39]
Marine Wing Support Squadron 372 [MWSS-372]
Marine Attack Squadron 231 [VMA-231]
Marine Medium Tiltrotor Squadron 261 [VMM-261]

Marine Logistics Combat Element

Combat Logistics Regiment 2 [CLR-2]
1st Dental Battalion [1st Den Bn]
7th Engineer Support Battalion [7th Eng Spt Bn]
Combat Logistics Battalion 6 [CLB-6]

I Marine Expeditionary Force (Forward) [I MEF (Fwd)]

Command Element

I MEF Headquarters Group
3d Battalion, 25th Marines [3d Bn, 25th Mar]
9th Communications Battalion [9th Comm Bn]
1st Radio Battalion [1st Radio Bn]
1st Air Naval Gunfire Liaison Company [1st ANGLICO]
2d Air Naval Gunfire Liaison Company [2d ANGLICO]
5th Air Naval Gunfire Liaison Company [5th ANGLICO]
3d Low Altitude Air Defense Battalion [3d LAAD BN]

Marine Ground Combat Element

- Regimental Combat Team 1 [RCT-1]
  - 2d Battalion, 1st Marines [2d Bn, 1st Mar]
  - 3d Battalion, 1st Marines [3d Bn, 1st Mar]
- Regimental Combat Team 2 [RCT-2]
  - 1st Battalion, 2d Marines [1st Bn, 2d Mar]
  - 2d Battalion, 2d Marines [2d Bn, 2d Mar]
  - 3d Battalion, 2d Marines [3d Bn, 2d Mar]
  - 3d Battalion, 9th Marines [3d Bn, 9th Mar]
- 1st Battalion, 3d Marines [1st Bn, 3d Mar]
- 2d Battalion, 3d Marines [2d Bn, 3d Mar]
- 3d Battalion, 3d Marines [3d Bn, 3d Mar]
- 1st Battalion, 5th Marines [1st Bn, 5th Mar]
- 3d Battalion, 5th Marines [3d Bn, 5th Mar]
- 1st Battalion, 6th Marines [1st Bn, 6th Mar]
- 2d Battalion, 6th Marines [2d Bn, 6th Mar]
- 3d Battalion, 6th Marines [3d Bn, 6th Mar]
- 2d Battalion, 9th Marines [2d Bn, 9th Mar]
- Regimental Combat Team 7 [RCT-7]
  - 3d Battalion, 7th Marines [3d Bn, 7th Mar]
  - 3d Battalion, 4th Marines [3d Bn, 4th Mar]
- 1st Battalion, 8th Marines [1st Bn, 8th Mar]
- 1st Battalion, 10th Marines [1st Bn, 10th Mar]
- 3d Battalion, 10th Marines [3d Bn, 10th Mar]
- 1st Battalion, 11th Marines [1st Bn, 11th Mar]
- 1st Light Armored Reconnaissance Battalion [1st LAR]
- 3d Light Armored Reconnaissance Battalion [3d LAR]
- 4th Light Armored Reconnaissance Battalion [4th LAR]
- 1st Reconnaissance Battalion [1st Recon]
- 2d Reconnaissance Battalion [2d Recon]
- 3d Reconnaissance Battalion [3d Recon]
- 1st Combat Engineer Battalion [1st CEB]
- 2d Combat Engineer Battalion [2d CEB]
- 3d Combat Engineer Battalion [3d CEB]

Marine Aviation Combat Element

- 3d Marine Aircraft Wing (Forward) [3d MAW (Fwd)]
  - Marine Air Control Group 38 (Forward) [MACG-38 (Fwd)]
  - Marine Heavy Helicopter Squadron 361 [HMH-361]

Marine Heavy Helicopter Squadron 362 [HMH-362]
Marine Light Attack Helicopter Squadron 169 [HMLA-169]
Marine Light Attack Helicopter Squadron 369 [HMLA-369]
Marine Fighter Attack Squadron 122 [VMFA-122]
Marine Fighter Attack Squadron 232 [VMFA-232]
Marine Aerial Refueler Transport Squadron 252 (-) [VMGR-252 (-)]
Marine Medium Tiltrotor Squadron 261 [VMM-261]
Marine Medium Tiltrotor Squadron 264 (Reinforced) [VMM-264 (Rein)]
Marine Medium Tiltrotor Squadron 365 [VMM-365]
Marine Unmanned Aerial Vehicle Squadron 1 [VMU-1]
Marine Unmanned Aerial Vehicle Squadron 2 [VMU-2]
Marine Aviation Logistics Squadron 16 [MALS-16]
Marine Wing Support Squadron 274 [MWSS-274]
Marine Wing Support Squadron 373 [MWSS-373]

Marine Logistics Combat Element

1st Marine Logistics Group (Forward) [1st MLG (Fwd)]
Combat Logistics Battalion 2 [CLB-2]
Combat Logistics Battalion 3 [CLB-3]
Combat Logistics Battalion 5 [CLB-5]
Combat Logistics Battalion 6 [CLB-6]
1st Maintenance Battalion [1st Maint Bn]
1st Supply Battalion [1st Sup Bn]
7th Engineer Support Battalion [7th Eng Spt Bn]
8th Engineer Support Battalion [8th Eng Spt Bn]
9th Engineer Support Battalion [9th Eng Spt Bn]
MAGTF Support Battalion [MAGTF Spt Bn]

2011

International Security Assistance Force, U.S. Forces-Afghanistan [ISAF USF-A]

Special Operations Task Force-West

1st Marine Special Operations Battalion [1st MSOB]
Marine Special Operations Company 811 (10 March 2011–30 October 2011)
Marine Special Operations Company 812 (29 September 2011–1 June 2012)
Marine Special Operations Company 813 (26 August 2010–April 2011)
2d Marine Special Operations Battalion [2d MSOB]

Marine Special Operations Task Force-West HQ (31 May 2010–1 February 2011)
Marine Special Operations Company 821 (18 November 2010–1 August 2011)
Marine Special Operations Company 822 (12 June 2011–1 March 2012)
Marine Special Operations Company 823 (30 April 2010–1 February 2011)

3d Marine Special Operations Battalion [3d MSOB]

Marine Special Operations Team 8311 to Marine Special Operations Company 811 (29 June 2011–15 February 2012)

9th Air and Space Expeditionary Task Force-Afghanistan [9th AETF-A]

Marine Tactical Electronic Warfare Squadron 1 [VMAQ-1]
Marine Tactical Electronic Warfare Squadron 3 [VMAQ-3]
Marine Tactical Electronic Warfare Squadron 4 [VMAQ-4]

International Security Assistance Force, Regional Command-Southwest [ISAF RC-SW]

I Marine Expeditionary Force (Forward) [I MEF (Fwd)]

Command Element

I MEF Headquarters Group

3d Battalion, 25th Marines [3d Bn, 25th Mar]
2d Radio Battalion (Forward) [2d Radio Bn (Fwd)]
8th Communication Battalion (Forward) [8th Comm Bn (Fwd)]
2d Air Naval Gunfire Liaison Company [2d ANGLICO]

Marine Ground Combat Element

1st Marine Division (Forward) [1st MarDiv (Fwd)]

Regimental Combat Team 1 [RCT-1]

2d Battalion, 1st Marines [2d Bn, 1st Mar]
2d Battalion, 3d Marines [2d Bn, 3d Mar]
2d Battalion, 8th Marines [2d Bn, 8th Mar]
2d Battalion, 9th Marines [2d Bn, 9th Mar]
3d Battalion, 9th Marines [3d Bn, 9th Mar]

Regimental Combat Team 2 [RCT-2]

3d Battalion, 5th Marines [3d Bn, 5th Mar]
1st Battalion, 8th Marines [1st Bn, 8th Mar]
3d Battalion, 8th Marines [3d Bn, 8th Mar]

3d Light Armored Reconnaissance Battalion [3d LAR]
2d Reconnaissance Battalion [2d Recon]

Marine Aviation Combat Element

3d Marine Aircraft Wing (Forward) [3d MAW (Fwd)]

Marine Air Control Group 38 (Forward) [MACG-38 (Fwd)]

Marine Heavy Helicopter Squadron 361 [HMH-361]

Marine Heavy Helicopter Squadron 362 [HMH-362]

Marine Heavy Helicopter Squadron 461 [HMH-461]

Marine Heavy Helicopter Squadron 463 [HMH-463]

Marine Light Attack Helicopter Squadron 169 [HMLA-169]

Marine Fighter Attack Squadron 122 [VMFA-122]

Marine Aerial Refueler Transport Squadron 252 (-) [VMGR-252 (-)]

Marine Medium Tiltrotor Squadron 264 [VMM-264]

Marine Medium Tiltrotor Squadron 266 (Reinforced) [VMM-266 (Rein)]

Marine Medium Tiltrotor Squadron 365 [VMM-365]

Marine Unmanned Aerial Vehicle Squadron 2 [VMU-2]

Marine Aviation Logistics Squadron 16 [MALS-16]

Marine Wing Support Squadron 373 [MWSS-373]

Marine Logistics Combat Element

1st Marine Logistics Group (Forward) [1st MLG (Fwd)]

Combat Logistics Battalion 2 [CLB-2]

Combat Logistics Battalion 3 [CLB-3]

Combat Logistics Battalion 8 [CLB-8]

1st Supply Battalion [1st Sup Bn]

8th Engineer Support Battalion [8th Eng Spt Bn]

II Marine Expeditionary Force (Forward) [II MEF (Fwd)]

Command Element

II MEF Headquarters Group

1st Battalion, 23d Marines [1st Bn, 23d Mar]

1st Battalion, 25th Marines [1st Bn, 25th Mar]

2d Intelligence Battalion (Forward) [2d Intel Bn (Fwd)]

2d Radio Battalion (Forward) [2d Radio Bn (Fwd)]

8th Communication Battalion (Forward) [8th Comm Bn (Fwd)]

1st Air Naval Gunfire Liaison Company [1st ANGLICO]

2d Air Naval Gunfire Liaison Company [2d ANGLICO]

Marine Ground Combat Element

2d Marine Division (Forward) [2d MarDiv (Fwd)]

Regimental Combat Team 1 [RCT-1]

2d Battalion, 1st Marines [2d Bn, 1st Mar]
1st Battalion, 3d Marines [1st Bn, 3d Mar]
2d Battalion, 3d Marines [2d Bn, 3d Mar]
1st Battalion, 6th Marines [1st Bn, 6th Mar]
3d Battalion, 6th Marines [3d Bn, 6th Mar]
2d Battalion, 8th Marines [2d Bn, 8th Mar]
1st Battalion, 9th Marines [1st Bn, 9th Mar]
3d Battalion, 9th Marines [3d Bn, 9th Mar]

Regimental Combat Team 5 [RCT-5]
1st Battalion, 3d Marines [1st Bn, 3d Mar]
3d Battalion, 3d Marines [3d Bn, 3d Mar]
2d Battalion, 6th Marines [2d Bn, 6th Mar]
3d Battalion, 6th Marines [3d Bn, 6th Mar]
1st Battalion, 9th Marines [1st Bn, 9th Mar]
2d Battalion, 9th Marines [2d Bn, 9th Mar]

Regimental Combat Team 8 [RCT-8]
3d Battalion, 2d Marines [3d Bn, 2d Mar]
2d Battalion, 4th Marines [2d Bn, 4th Mar]
3d Battalion, 4th Marines [3d Bn, 4th Mar]
1st Battalion, 5th Marines [1st Bn, 5th Mar]
3d Battalion, 5th Marines [3d Bn, 5th Mar]
1st Battalion, 6th Marines [1st Bn, 6th Mar]
3d Battalion, 7th Marines [3d Bn, 7th Mar]
1st Battalion, 8th Marines [1st Bn, 8th Mar]
3d Battalion, 8th Marines [3d Bn, 8th Mar]

1st Battalion, 10th Marines [1st Bn, 10th Mar]
1st Battalion, 12th Marines [1st Bn, 12th Mar]
2d Combat Engineer Battalion [2d CEB]
3d Combat Engineer Battalion [3d CEB]
1st Light Armored Reconnaissance Battalion [1st LAR]
3d Light Armored Reconnaissance Battalion [3d LAR]
2d Reconnaissance Battalion [2d Recon]
3d Reconnaissance Battalion [3d Recon]

Marine Aviation Combat Element

2d Marine Aircraft Wing (Forward) [2d MAW (Fwd)]
Marine Air Control Group 28 (-) [MACG-28 (-)]
Marine Heavy Helicopter Squadron 363 [HMH-363]
Marine Heavy Helicopter Squadron 463 [HMH-463]
Marine Heavy Helicopter Squadron 464 [HMH-464]

Marine Light Attack Helicopter Squadron 169 [HMLA-169]
Marine Light Attack Helicopter Squadron 267 [HMLA-267]
Marine Light Attack Helicopter Squadron 369 (-) [HMLA-369 (-)]
Marine Attack Squadron 223 [VMA-223]
Marine Attack Squadron 513 [VMA-513]
Marine Fighter Attack Squadron 122 [VMFA-122]
Marine Fighter Attack Squadron 513 [VMFA-513]
Marine Aerial Refueler Transport Squadron 252 (-) [VMGR-252 (-)]
Marine Medium Tiltrotor Squadron 162 [VMM-162]
Marine Medium Tiltrotor Squadron 264 [VMM-264]
Marine Medium Tiltrotor Squadron 269 [VMM-269]
Marine Unmanned Aerial Vehicle Squadron 1 [VMU-1]
Marine Unmanned Aerial Vehicle Squadron 2 [VMU-2]
Marine Unmanned Aerial Vehicle Squadron 3 [VMU-3]
Marine Aviation Logistics Squadron 40 [MALS-40]
Marine Wing Support Squadron 272 [MWSS-272]

Marine Logistics Combat Element

2d Marine Logistics Group (Forward) [2d MLG (Fwd)]
Combat Logistics Battalion 1 [CLB-1]
Combat Logistics Battalion 3 [CLB-3]
Combat Logistics Battalion 6 [CLB-6]
Combat Logistics Battalion 7 [CLB-7]
Combat Logistics Battalion 8 [CLB-8]
7th Engineer Support Battalion [7th Eng Spt Bn]
8th Engineer Support Battalion [8th Eng Spt Bn]
9th Engineer Support Battalion [9th Eng Spt Bn]
MAGTF Support Battalion [MAGTF Spt Bn]

2012

International Security Assistance Force, U.S. Forces-Afghanistan [ISAF USF-A]

Special Operations Task Force-West

1st Marine Special Operations Battalion [1st MSOB]
Special Operations Task Force 81.2 (4 January 2012–1 October 2012)
Marine Special Operations Company 811 (11 November 2012–7 July 2013)
Marine Special Operations Company 812 (29 September 2011–1 June 2012)

Marine Special Operations Company 813 (7 June 2012–6 February 2013)

2d Marine Special Operations Battalion [2d MSOB]

Special Operations Task Force 82.2 HQ (1 October 2012–1 June 2013)

Marine Special Operations Company 821 (15 September 2012–15 April 2013)

Marine Special Operations Company 822 (12 June 2011–1 March 2012)

Marine Special Operations Company 823 (16 June 2012–15 August 2012)

3d Marine Special Operations Battalion [3d MSOB]

Marine Special Operations Team 8311 to Marine Special Operations Company 811 (29 June 2011–15 February 2012)

Marine Special Operations Team 8323 to Marine Special Operations Company 821 (13 August 2012–15 May 2013)

9th Air and Space Expeditionary Task Force-Afghanistan [9th AETF-A]

Marine Tactical Electronic Warfare Squadron 1 [VMAQ-1]

Marine Tactical Electronic Warfare Squadron 2 [VMAQ-2]

Marine Tactical Electronic Warfare Squadron 3 [VMAQ-3]

International Security Assistance Force, Regional Command-Southwest [ISAF RC-SW]

II Marine Expeditionary Force (Forward) [II MEF (Fwd)]

Command Element

II MEF Headquarters Group

1st Battalion, 25th Marines [1st Bn, 25th Mar]

2d Intelligence Battalion [2d Intel Bn]

2d Radio Battalion [2d Radio Bn]

8th Communication Battalion [8th Comm Bn]

5th Air Naval Gunfire Liaison Company [5th ANGLICO]

Marine Ground Combat Element

2d Marine Division (Forward) [2d MarDiv (Fwd)]

Regimental Combat Team 5 [RCT-5]

3d Battalion, 3d Marines [3d Bn, 3d Mar]

2d Battalion, 6th Marines [2d Bn, 6th Mar]

2d Battalion, 9th Marines [2d Bn, 9th Mar]

Regimental Combat Team 6 [RCT-6]

2d Battalion, 4th Marines [2d Bn, 4th Mar]

3d Battalion, 7th Marines [3d Bn, 7th Mar]

1st Battalion, 8th Marines [1st Bn, 8th Mar]

Regimental Combat Team 8 [RCT-8]
2d Battalion, 4th Marines [2d Bn, 4th Mar]
1st Battalion, 6th Marines [1st Bn, 6th Mar]
3d Battalion, 7th Marines [3d Bn, 7th Mar]
2d Battalion, 11th Marines [2d Bn, 11th Mar]
3d Combat Engineer Battalion [3d CEB]
1st Light Armored Reconnaissance Battalion [1st LAR]
1st Reconnaissance Battalion [1st Recon Bn]

Marine Aviation Combat Element

2d Marine Aircraft Wing (Forward) [2d MAW (Fwd)]
Marine Air Control Group 28 (-) [MACG-28 (-)]
Marine Heavy Helicopter Squadron 363 [HMH-363]
Marine Heavy Helicopter Squadron 464 [HMH-464]
Marine Light Attack Helicopter Squadron 369 [HMLA-369]
Marine Attack Squadron 223 [VMA-223]
Marine Medium Tiltrotor Squadron 269 [VMM-269]
Marine Medium Tiltrotor Squadron 365 [VMM-365]
Marine Unmanned Aerial Vehicle Squadron 1 [VMU-1]
Marine Aviation Logistics Squadron 40 (Forward) [MALS-40 (Fwd)]
Marine Wing Support Squadron 371 [MWSS-371]

Marine Logistics Combat Element

1st Marine Logistics Group (Forward) [1st MLG (Fwd)]
Combat Logistics Battalion 1 [CLB-1]
Combat Logistics Battalion 4 [CLB-4]
9th Engineer Support Battalion [9th Eng Spt Bn]
MAGTF Support Battalion [MAGTF Spt Bn]

I Marine Expeditionary Force (Forward) [I MEF (Fwd)]

Command Element

I MEF Headquarters Group
1st Intelligence Battalion [1st Intel Bn]
1st Radio Battalion [1st Radio Bn]
9th Communication Battalion [9th Comm Bn]

Marine Ground Combat Element

1st Marine Division (Forward) [1st MarDiv (Fwd)]
Regimental Combat Team 5 [RCT-5]

3d Battalion, 3d Marines [3d Bn, 3d Mar]
2d Battalion, 6th Marines [2d Bn, 6th Mar]
2d Battalion, 9th Marines [2d Bn, 9th Mar]
Regimental Combat Team 6 [RCT-6]
1st Battalion, 1st Marines [1st Bn, 1st Mar]
2d Battalion, 5th Marines [2d Bn, 5th Mar]
1st Battalion, 7th Marines [1st Bn, 7th Mar]
2d Battalion, 7th Marines [2d Bn, 7th Mar]
3d Battalion, 7th Marines [3d Bn, 7th Mar]
1st Battalion, 8th Marines [1st Bn, 8th Mar]
3d Battalion, 8th Marines [3d Bn, 8th Mar]
Regimental Combat Team 7 [RCT-7]
1st Battalion, 1st Marines [1st Bn, 1st Mar]
2d Battalion, 7th Marines [2d Bn, 7th Mar]
3d Battalion, 8th Marines [3d Bn, 8th Mar]
3d Battalion, 9th Marines [3d Bn, 9th Mar]
2d Battalion, 10th Marines [2d Bn, 10th Mar]
2d Battalion, 11th Marines [2d Bn, 11th Mar]
1st Combat Engineer Battalion [1st CEB]
3d Combat Engineer Battalion [3d CEB]
1st Light Armored Reconnaissance Battalion [1st LAR]
3d Light Armored Reconnaissance Battalion [3d LAR]
1st Reconnaissance Battalion [1st Recon Bn]

Marine Aviation Combat Element

3d Marine Aircraft Wing (Forward) [3d MAW (Fwd)]
Marine Air Control Group 38 (-) (Forward) [MACG-38 (-) (Fwd)]
Marine Heavy Helicopter Squadron 361 [HMH-361]
Marine Heavy Helicopter Squadron 362 [HMH-362]
Marine Heavy Helicopter Squadron 363 [HMH-363]
Marine Heavy Helicopter Squadron 466 [HMH-466]
Marine Light Attack Helicopter Squadron 169 [HMLA-169]
Marine Light Attack Helicopter Squadron 369 [HMLA-369]
Marine Light Attack Helicopter Squadron 469 [HMLA-469]
Marine Attack Squadron 211 [VMA-211]
Marine Attack Squadron 223 [VMA-223]
Marine Attack Squadron 231 [VMA-231]
Marine Aerial Refueler Transport Squadron 352 (-) [VMGR-352 (-)]
Marine Medium Tiltrotor Squadron 161 [VMM-161]
Marine Medium Tiltrotor Squadron 365 [VMM-365]

Marine Unmanned Aerial Vehicle Squadron 1 [VMU-1]
Marine Unmanned Aerial Vehicle Squadron 2 [VMU-2]
Marine Unmanned Aerial Vehicle Squadron 3 [VMU-3]
Marine Aviation Logistics Squadron 16 (Forward) [MALS-16 (Fwd)]
Marine Wing Support Squadron 273 [MWSS-273]
Marine Wing Support Squadron 373 [MWSS-373]

Marine Logistics Combat Element

1st Marine Logistics Group (Forward) [1st MLG (Fwd)]
Combat Logistics Battalion 1 [CLB-1]
Combat Logistics Battalion 2 [CLB-2]
Combat Logistics Battalion 4 [CLB-4]
Combat Logistics Battalion 5 [CLB-5]
1st Maintenance Battalion (-) (Reinforced) [1st Mnt Bn (-) (Rein)]
8th Engineer Support Battalion [8th Eng Spt Bn]
9th Engineer Support Battalion [9th Eng Spt Bn]

2013

International Security Assistance Force, U.S. Forces-Afghanistan [ISAF USF-A]

Special Operations Task Force-West

1st Marine Special Operations Battalion [1st MSOB]
Marine Special Operations Company 811 (11 November 2012–7 July 2013)
Marine Special Operations Company 812 (7 July 2013–7 February 2014)
Marine Special Operations Company 813 (7 June 2012–6 February 2013)
2d Marine Special Operations Battalion [2d MSOB]
Special Operations Task Force 82.2 (1 October 2012–1 June 2013
Marine Special Operations Company 821 (15 September 2012–15 May 2013)
Marine Special Operations Company 822 (15 April 2013–15 November 2013)
Marine Special Operations Company 823 (7 November 2013–1 June 2014)
3d Marine Special Operations Battalion [3d MSOB]
Marine Special Operations Team 8323 (13 August 2012–15 May 2013)

9th Air and Space Expeditionary Task Force-Afghanistan [9th AETF-A]

Marine Tactical Electronic Warfare Squadron 3 [VMAQ-3]
Marine Tactical Electronic Warfare Squadron 4 [VMAQ-4]

International Security Assistance Force, Regional Command-Southwest [ISAF RC-SW]

I Marine Expeditionary Force (Forward) [I MEF (Fwd)]

Command Element

- I MEF Headquarters Group
  - 1st Intelligence Battalion [1st Intel Bn]
  - 1st Radio Battalion [1st Radio Bn]
  - 9th Communication Battalion [9th Comm Bn]

Marine Ground Combat Element

- Regimental Combat Team 7 [RCT-7]
  - 2d Battalion, 7th Marines [2d Bn, 7th Mar]
  - 3d Battalion, 9th Marines [3d Bn, 9th Mar]
- 2d Battalion, 10th Marines [2d Bn, 10th Mar]
- 2d Combat Engineer Battalion [2d CEB]

Marine Aviation Combat Element

- 3d Marine Aircraft Wing (Forward) [3d MAW (Fwd)]
  - Marine Air Control Group 28 (Forward) [MACG-28 (Fwd)]
  - Marine Heavy Helicopter Squadron 361 [HMH-361]
  - Marine Heavy Helicopter Squadron 461 [HMH-461]
  - Marine Light Attack Helicopter Squadron 169 [HMLA-169]
  - Marine Attack Squadron 211 [VMA-211]
  - Marine Attack Squadron 231 [VMA-231]
  - Marine Aerial Refueler Transport Squadron 252 (-) [VMGR-252 (-)]
  - Marine Medium Tiltrotor Squadron 161 [VMM-161]
  - Marine Unmanned Aerial Vehicle Squadron 3 (Forward) [VMU-3 (Fwd)]
  - Marine Aviation Logistics Squadron 40 [MALS-40]
  - Marine Wing Support Squadron 373 (Forward) [MWSS-373 (Fwd)]

Marine Logistics Combat Element

- Combat Logistics Battalion 2 [CLB-2]

II Marine Expeditionary Force (Forward) [II MEF (Fwd)]

Command Element

- II MEF Headquarters Group
  - 2d Law Enforcement Battalion [2d LEB]

2d Radio Battalion [2d Radio Bn]
8th Communications Battalion [8th Comm Bn]
2d Air Naval Gunfire Liaison Company [2d ANGLICO]

Marine Ground Combat Element

2d Marine Regiment [2d Mar]
Regimental Combat Team 7 [RCT-7]
3d Battalion, 4th Marines [3d Bn, 4th Mar]
2d Battalion, 7th Marines [2d Bn, 7th Mar]
3d Battalion, 7th Marines [3d Bn, 7th Mar]
2d Battalion, 8th Marines [2d Bn, 8th Mar]
3d Battalion, 9th Marines [3d Bn, 9th Mar]
1st Combat Engineer Battalion (-) (Reinforced) [1st CEB (-) (Rein)]
2d Combat Engineer Battalion [2d CEB]
3d Combat Engineer Battalion (Forward) [3d CEB (Fwd)]

Marine Aviation Combat Element

2d Marine Aircraft Wing (Forward) [2d MAW (Fwd)]
Marine Air Control Group 28 (Forward) [MACG-28 (Fwd)]
Marine Heavy Helicopter Squadron 361 [HMH-361]
Marine Heavy Helicopter Squadron 461 [HMH-461]
Marine Heavy Helicopter Squadron 462 [HMH-462]
Marine Light Attack Helicopter Squadron 167 [HMLA-167]
Marine Light Attack Helicopter Squadron 169 [HMLA-169]
Marine Light Attack Helicopter Squadron 369 [HMLA-369]
Marine Attack Squadron 231 [VMA-231]
Marine Attack Squadron 311 [VMA-311]
Marine Aerial Refueler Transport Squadron 252 (-) [VMGR-252 (-)]
Marine Medium Tiltrotor Squadron 161 [VMM-161]
Marine Medium Tiltrotor Squadron 165 [VMM-165]
Marine Medium Tiltrotor Squadron 264 [VMM-264]
Marine Unmanned Aerial Vehicle Squadron 2 (Forward) [VMU-2 (Fwd)]
Marine Unmanned Aerial Vehicle Squadron 3 (Forward) [VMU-3 (Fwd)]
Marine Aviation Logistics Squadron 40 [MALS-40]
Marine Wing Support Squadron 271 (-) (Forward) [MWSS-271 (-) (Fwd)]
Marine Wing Support Squadron 372 (Forward) [MWSS 372 (Fwd)]

Marine Logistics Combat Element

Combat Logistics Regiment 2 [CLR-2]
Combat Logistics Battalion 2 [CLB-2]

2014

International Security Assistance Force, U.S. Forces-Afghanistan [ISAF USF-A]

Special Operations Task Force-West

- 1st Marine Special Operations Battalion [1st MSOB]
    - Special Operations Task Force 81.3 HQ (1 February–1 October 2014)
- 2d Marine Special Operations Battalion [2d MSOB]
    - Marine Special Operations Company 821 (16 May 2014–15 September 2014)
    - Marine Special Operations Company 823 (7 November 2013–1 June 2014)

9th Air and Space Expeditionary Task Force-Afghanistan [9th AETF-A]

- Marine Tactical Electronic Warfare Squadron 4 [VMAQ-4]

International Security Assistance Force, Regional Command-Southwest [ISAF RC-SW]

II Marine Expeditionary Force (Forward) [II MEF (Fwd)]

Command Element

- II MEF Headquarters Group
    - 2d Law Enforcement Battalion [2d LEB]
    - 2d Radio Battalion [2d Radio Bn]
    - 8th Communication Battalion (Forward) [8th Comm Bn]
    - 2d Air Naval Gunfire Liaison Company [2d ANGLICO]

Marine Ground Combat Element

- 2d Marine Regiment [2d Mar]
    - 3d Battalion, 7th Marines [3d Bn, 7th Mar]
    - 1st Battalion, 9th Marines [1st Bn, 9th Mar]
- 1st Combat Engineer Battalion (-) (Reinforced) [1st CEB (-) (Rein)]

Marine Aviation Combat Element

- 2d Marine Aircraft Wing (Forward) [2d MAW (Fwd)]
    - Marine Air Control Group 38 (-) [MACG-38 (-)]
    - Marine Heavy Helicopter Squadron 462 [HMH-462]
    - Marine Light Attack Helicopter Squadron 369 [HMLA-369]
    - Marine Attack Squadron 311 [VMA-311]
    - Marine Aerial Refueler Transport Squadron 352 (-) [VMGR-352 (-)]

Marine Medium Tiltrotor Squadron 165 [VMM-165]
Marine Unmanned Aerial Vehicle Squadron 2 (-) [VMU-2 (-)]
Marine Air Logistics Squadron 70 [MALS-70]
Marine Wing Support Squadron 372 (Forward) [MWSS-372 (Fwd)]

Marine Logistics Combat Element

Combat Logistics Battalion 6 [CLB-6]
Combat Logistics Battalion 7 [CLB-7]

Marine Expeditionary Brigade-Afghanistan [MEB-A]

Command Element

MEB-A Headquarters Group
1st Radio Battalion (Forward) [1st Radio Bn (Fwd)]
9th Communication Battalion (Forward) [9th Comm Bn (Fwd)]
1st Air Naval Gunfire Liaison Company [1st ANGLICO]

Marine Ground Combat Element

1st Marine Regiment [1st Mar]
1st Battalion, 2d Marines [1st Bn, 2d Mar]
1st Battalion, 7th Marines [1st Bn, 7th Mar]
3d Battalion, 7th Marines [3d Bn, 7th Mar]
1st Battalion, 9th Marines [1st Bn, 9th Mar]
1st Combat Engineer Battalion (-) (Reinforced) [1st CEB (-) (Rein)]
2d Combat Engineer Battalion [2d CEB]

Marine Aviation Combat Element

Marine Aircraft Group 26 [MAG-26]
Marine Air Control Group 38 (-) [MACG-38 (-)]
Marine Heavy Helicopter Squadron 462 [HMH-462]
Marine Heavy Helicopter Squadron 466 [HMH-466]
Marine Light Attack Helicopter Squadron 369 [HMLA-369]
Marine Light Attack Helicopter Squadron 467 [HMLA-467]
Marine Attack Squadron 311 [VMA-311]
Marine Aerial Refueler Transport Squadron 352 (-) [VMGR-352 (-)]
Marine Medium Tiltrotor Squadron 261 [VMM-261]
Marine Unmanned Aerial Vehicle Squadron 1 (-) [VMU-1 (-)]
Marine Unmanned Aerial Vehicle Squadron 2 (-) [VMU-2 (-)]
Marine Air Logistics Squadron 70 [MALS-70]

Marine Wing Support Squadron 274 (Forward) [MWSS-274 (Fwd)]
Marine Wing Support Squadron 372 (Forward) [MWSS-372 (Fwd)]

Marine Logistics Combat Element

Combat Logistics Battalion 7 [CLB-7]

Appendix C: Selected Terms and Abbreviations from Operation Enduring Freedom

ABP	Afghan border police
ABV	(M1) Assault Breacher Vehicle
ACE	Aviation Combat Element
al-Qaeda	"The base"; an international militant Islamic terrorist organization
ANA	Afghan National Army
ANGLICO	Air Naval Gunfire Liaison Company
ANP	Afghan National Police
ANSF	Afghan National Security Forces
AOR	Area of responsibility
ArCent	U.S. Army Forces Central Command
ARG	Amphibious Ready Group
ASNF	Afghan Special Narcotics Force
AT	Antiterrorism
AUP	Afghan Uniformed Police
BLT	Battalion Landing Team
burka	A loose enveloping garment worn by some Muslim women to cloak their faces and bodies in public places
CAAT	Combined Antiarmor Team
CAG	Civil Affairs Group
CAS	Close Air Support
CATF	Commander Amphibious Task Force
CEB	Combat Engineer Battalion
CENTCOM	U.S. Central Command
CERP	Commander's Emergency Response Program
CFLCC	Coalition Forces Land Component Command
CFACC	Coalition Forces Air Component Commander
CFMCC	Coalition Forces Maritime Component Command
CFSOCC	Coalition Forces Special Operations Component Command
CG	Commanding general
CJSOTF	Combined Joint Special Operations Task Force

CJTF	Combined Joint Task Force
CLB	Combat Logistics Battalion
CLF	Commander Landing Force
CLR	Combat Logistics Regiment
CMOC	Civil-Military Operations Center
COP	Combat outpost
COIN	Counterinsurgency
ComdC	Command chronology
ConPlan	Concept of operations plan
CSAR	Combat Search and Rescue
CSTC	Combined Security Transition Command
CVN	Aircraft carrier, nuclear (fixed-wing aircraft)
CVW	Carrier Air Wing
D-Day	Day on which operations are scheduled to commence
DASC	Direct Air Support Center
DOD	Department of Defense
DON	Department of the Navy
EODMU	Explosive Ordnance Disposal Mobile Unit
ESB	Engineer Support Battalion
ETT	Embedded Training Team
FAC	Forward Air Controller
FAC(A)	Forward Air Controller (Airborne)
FARP	Forward arming and refueling point
fatwa	Islamic religious ruling
FDC	Fire direction center
FET	Female Engagement Team
FOB	Forward operating base
GBOSS	Ground-based operational surveillance system
GIRoA	Government of the Islamic Republic of Afghanistan
GMTI	Ground moving target indicator
GRC	Gray Research Center
H-Hour	Hour when operation is scheduled to commence
HESCO	A multicellular wall system manufactured from wire mesh and lined with a heavy-duty material then filled with earth to provide a semi-permanent barrier for military installations

HIMARS	High Mobility Artillery Rocket System
HMH	Marine Heavy Helicopter Squadron
HMLA	Marine Light Attack Helicopter Squadron
HMM	Marine Medium Helicopter Squadron
HMMWV	High Mobility Multipurpose Wheeled Vehicle or humvee
HQMC	Headquarters Marine Corps
IED	Improvised explosive device
IO	Information operations
IJC	ISAF Joint Command
ISAF	International Security Assistance Force
ISCI	Interim security for critical infrastructure
ISI	The Directorate for Inter-Services Intelligence (Pakistan)
ISR	Intelligence, surveillance, and reconnaissance
JCS	Joint Chiefs of Staff
JDAM	Joint direct attack munition
JIC	Joint Intelligence Center
jihad	Holy war waged as an Islamic religious duty
JPEL	Joint Prioritized Effects List
JSTARS	Joint surveillance target attack radar system
kandak	Afghan National Army Battalion
LAAD	Low altitude air defense
LAR	Light armored reconnaissance battalion
LAV	Light armored vehicle
LTD	Laser target designator
LHA	Landing helicopter assault (general purpose amphibious assault ship)
LHD	Landing helicopter dock (multipurpose amphibious assault ship)
LD	Line of departure
loya jirga	Traditional meeting of tribal elders to solve problems
LPD	Landing platform dock (amphibious transport dock)
LPH	Landing platform helicopter (amphibious assault ship)
LSA	Logistics support area
LSD	Landing Ship, Dock (amphibious assault ship)
LVS	Logistics vehicle system
MACG	Marine air control group
MACS	Marine air control squadron

madrassa	Islamic religious school
MAG	Marine aircraft group
MAGTF	Marine air-ground task force
MALS	Marine aviation logistics squadron
MarCent	U.S. Marine Corps Forces Central Command
MARSOC	Marine Corps Forces Special Operations Command
MARCORSYSCOM	Marine Corps Systems Command
MAW	Marine aircraft wing
MCCDC	Marine Corps Combat Development Command
MCHD	Marine Corps History Division
MEB	Marine expeditionary brigade
MEF	Marine expeditionary force
MEU	Marine expeditionary unit
MICLIC	Mine-clearing line charge
MLG	Marine logistics group
MPF	Maritime prepositioning force
MRAP	Mine-resistant, ambush protected
MSOB	Marine special operations battalion
MSOC	Marine special operations company
MSOT	Marine special operations team
MSSG	Marine expeditionary unit (MEU) service support group
MTVR	Medium tactical vehicle replacement
mujahideen	Those who wage a jihad; holy warriors
mullah	Male religious teacher or leader who is schooled in Islamic law
MWHS	Marine wing headquarters squadron
MWSS	Marine wing support squadron
NATO	North Atlantic Treaty Organization
NavCent	U.S. Naval Forces Central Command
nm	Nautical mile
NMCB	Naval mobile construction battalion
ODA	Operational detachment alpha
OEF	Operation Enduring Freedom
OIF	Operation Iraqi Freedom
OP	Observation post
OSC–A	Office of Security Cooperation–Afghanistan
PA	Public affairs
Pashtunwali	A way of life and system of customary laws that stress honor above all else

PB	Patrol base
PR	Public relations
PRT	Provincial reconstruction team
QRF	Quick reaction force
RC	Regional command
RC-Southwest	Regional Command-Southwest
RCT	Regimental combat team
RIP	Relief in place
RPG	Rocket-propelled grenade
Sangar	Used by British forces to describe a watchtower
SASR	Special application scoped rifle
SAW	Squad Automatic Weapon
Seabee	U.S. Navy construction battalion
SEAL	Sea, air, and land (U.S. Navy Special Operations force)
SFAAT	Security Force Assistance Advisor Team
sharia	Islamic law
shura	A town hall-style council of decision making by consultation and deliberation
SIPRNET	Secret Internet protocol router network
SOC	Special operations capable
SOF	Special operations forces (generic)
SPMAGTF	Special purpose Marine air-ground task force
SORTS	Status of resources and training system
SWAN	Support wide area network
Taliban	"Seekers, religious students"; an Islamic-based, Afghan political military organization that emerged during 1994 and that ruled large parts of Afghanistan from 1996 to 2001
TF	Task force
TIC	Troops in contact (with the enemy)
TTF	Theater tactical force
TOA	Transfer of authority
TOW	Tube-launched, optically tracked, wire-guided missile
UAV	Unmanned aerial vehicle
UN	United Nations
UNAMA	United Nations Assistance Mission in Afghanistan
UNSC	United Nations Security Council

USA	United States Army
USAF	United States Air Force
USAID	United States Agency for International Development
USF-A	United States Forces-Afghanistan
USMC	United States Marine Corps
USN	United States Navy
USNS	United States naval ship
USO	United Service Organizations
USS	United States ship
USSOCOM	United States Special Operations Command
VMA	Marinc attack squadron
VMAQ	Marine tactical electronic warfare squadron
VMFA	Marine fighter attack squadron
VMFA(AW)	Marine all-weather fighter attack squadron
VMGR	Marine aerial refueler transport squadron
VMM	Marine medium tiltrotor squadron
VMU	Marine unmanned aerial vehicle squadron
wadi	A valley, ravine, or channel that is dry except in the rainy season

Appendix D:
Chronology of Events, 2010–14

2010

13 February	Operation Moshtarak, an offensive on the Afghan town of Marjah meant to eliminate the last Taliban in central Helmand Province, begins. This offensive is, up to this point, the largest joint operation in the war in Afghanistan. It is heavily publicized by ISAF in an attempt to demonstrate improvements in the Afghan National Army (ANA) and is meant to assess whether Afghan forces are capable of keeping peace in their country.
12 April	The 2d Marine Expeditionary Brigade (a.k.a. Marine Expeditionary Brigade–Afghanistan [MEB-A]) transfers authority to I Marine Expeditionary Force (Forward) (I MEF [Fwd]).
12 June	Operation New Dawn, an extension of Operation Moshtarak, is launched by International Security Assistance Force (ISAF) and ANA forces with the goal of forcing insurgents from areas between Marjah and Nawa in Helmand Province.
23 June	Gen Stanley A. McChrystal, USA, is relieved of his post as commander of U.S. forces in Afghanistan as a result of a controversial article in *Rolling Stone* in which he criticized the Obama administration. President Obama nominates Gen David H. Petraeus, USA, to replace McChrystal.
18 September	The Afghan parliamentary election is held to elect members to the Wolesi Jirga (or House of the People).
20 September	Northern Helmand Province is transferred from British control to that of the U.S. Marines. The 3d Battalion, 7th Marines, is the first Marine unit to replace British troops, deploying to Sangin, Helmand Province, and engaging Taliban insurgents.
October	The 3d Battalion, 5th Marines, relieves 3d Battalion, 7th Marines, in Sangin. The battalion immediately begins conducting aggressive operations against Taliban insurgents.
November	Representatives from North Atlantic Treaty Organization (NATO) countries

sign an agreement to return responsibility for security in Afghanistan to Afghan forces by the end of 2014. This declaration concurs with the planned drawdown of U.S. troops in Afghanistan, and schedules a transition process between American forces and Afghan security forces to begin in July 2011.

21 November — During a firefight with enemy insurgents in a small village in Marjah District, Helmand Province, LCpl W. Kyle Carpenter shields a fellow Marine from the detonation of a hand grenade, shielding the Marine from the blast but enduring severe wounds. For his courage, Carpenter is awarded the Medal of Honor.

27 December — The 3d Battalion, 9th Marines, relieves 2d Battalion, 6th Marines, in Marjah.

28–29 December — Marines of 2d Battalion, 3d Marines, conduct Operation East River, a partnered rehearsal drill that reveals the capabilities of ANA and Afghan Uniformed Police (AUP) forces in Nawa District, Helmand Province.

2011

January — Marine Special Operations Charlie Company works to establish two village stability operation positions in the Upper Gereshk Valley in Helmand Province and northern Farah Province. Multiple operations conducted by these Marines aim to decrease Taliban insurgents' freedom of movement, as well as increasing the local populace's freedom of movement and improving their connection to district and provincial governments.

6 January — The Pentagon announces that it will send 1,400 Marines to southern Afghanistan to build on security gains there.

14 January — Marines from 2d Battalion, 1st Marines, and ANA forces conduct Operation Godfather, an attack on the Taliban stronghold at Durzay, which is one of the last remaining strongholds in Garmsir District, Helmand Province. This operation concludes major fighting in Garmsir.

16–20 January — Marines of 3d Battalion, 9th Marines, conduct Operation Jalut (Goliath), in which more than 300 Marines and Afghan National Security Forces (ANSF) mass against insurgent forces in Marjah. Marine engineers quickly build a bridge and road in support of interim security for critical infrastructure (ISCI).

February–March — Focusing on the drug harvest of the poppy fields in Southern Helmand, Regimental Combat Team 1 (RCT-1) targets the processing centers and transportation assets used to produce heroin from the poppy crop.

1 February	Marine Heavy Helicopter Squadron 461 (HMH-461) relieves HMH-361 (-) at Camp Bastion in Helmand Province.
6 February	RCT-8 relieves RCT-2 at Forward Operating Base (FOB) Delaram in Delaram District.
8 February	The 2d Battalion, 8th Marines, relieves 2d Battalion, 9th Marines, in northern Marjah.
16 February	Elements of 3d Battalion, 9th Marines, conduct Operation Sulaiman in coordination with ANSF and Marines from 2d Battalion, 3d Marines. This action prevents the enemy from conducting operations in support of a spring offensive.
17 February	Marines of 3d Battalion, 5th Marines, conduct Operation Savage Wing, which clears suspected insurgent fighting, storage, and travel lanes near Camp Delaram 2. The Marines rely on indirect fire and supporting assets from 1st Battalion, 10th Marines, in execution of this operation.
17–20 February	Marines of 3d Battalion, 5th Marines, conduct Operation Red Hawk to disrupt enemy caches and the improvised explosive device (IED) production and emplacement network. This operation turns out multiple weapons caches and enemy activities, resulting in numerous small-arms engagements and enemy casualties.
21–24 February	Marines of 2d Battalion, 8th Marines, execute Operation Steel Curtain, which clears Barakzai Village in southeastern Trek Nawa and establishes new forward patrol bases from which Golf Company conducts counterinsurgency operations.
March	Marine Special Operations Alpha Company relieves Marine Special Operations Charlie Company in Afghanistan.
9 March	The 3d Marine Aircraft Wing (Forward) (3d MAW [Fwd]) conducts a transfer of authority with 2d MAW (Fwd) at Camp Leatherneck in Helmand Province.
10 March	HMH-463 relieves HMH-362 at Camp Bastion in Helmand Province.
16 March	The 1st MarDiv (Fwd) conducts a transfer of authority with 2d MarDiv (Fwd) at Camp Leatherneck.
	Marines of 3d Light Armored Reconnaissance Battalion (3d LAR) begin Operation Rawhide II in the southern Helmand River valley, where interdiction succeeds

in denying weapons, narcotics, and trafficking routes to the enemy insurgency. The operation continues until the battalion is relieved in May.

17 March — The 3d Battalion, 2d Marines, relieves 1st Battalion, 8th Marines, in Musa Qala and Now Zad Districts, northern Helmand Province.

21–30 March — Elements of RCT-1 execute Operation Watchtower II, a Coalition-led action meant to disrupt the enemy's abilities to conduct a spring offensive. Coalition and ANSF troops work together to conduct multiple targeted clearing and disruption missions, which increases the ANSF's ability to operate with minimal mentorship or supervision.

26 March — The II MEF (Fwd) assumed command of Regional Command-Southwest (RC-Southwest).

11 April — The 1st Battalion, 5th Marines, relieves 3d Battalion, 5th Marines, in Sangin District.

23 April–9 May — Elements of RCT-1 conduct Operation Harvest Moon in Helmand Province. This action disrupts enemy attempts to finish the poppy harvest, hinders the final preparations for the enemy forces' summer offensive, and facilitates the capture of key enemy personnel.

1 May — Osama bin Laden is killed during Operation Neptune Spear, a U.S. Navy SEAL raid on his compound in Abbottabad, Pakistan, nine-and-a-half years after his role in orchestrating the 11 September 2001 terrorist attacks on the United States. His death prompts U.S. policy makers to reevaluate the conduct of the war, as President Obama is scheduled to announce in June 2011 the withdrawal of the American surge troops sent to Afghanistan in December 2009.

6 May — The 1st Battalion, 3d Marines, relieves 2d Battalion, 1st Marines, in Garmsir District.

7 May — Marine Unmanned Aerial Vehicle Squadron 3 (VMU-3) relieves VMU-2 in Helmand Province.

15–17 May — Marines of 3d Battalion, 2d Marines, conduct Operation Talai Ghar (Keyhole), which denies insurgents sanctuary in the Salaam Bazaar area of southern Now Zad District.

19 May	The 1st Battalion, 12th Marines (-) (Rein), relieves 1st Battalion, 10th Marines (-) (Rein), to provide artillery support to the 1st MarDiv in Helmand Province.
21–24 May	Elements of 1st Battalion, 3d Marines, conduct Operation Audible, dispatching Alpha Company and a reinforced combined antiarmor team (CAAT) platoon to the Kartaka and Bartaka areas west of the Helmand River to engage insurgents there.
22 May	The 2d LAR relieves 3d LAR at FOB Payne in Helmand Province and continues conducting Operation Rawhide II in the area.
23 May	Marine Attack Squadron 513 (VMA-513) relieves VMA-122 at Kandahar Airfield in Afghanistan, after the two squadrons fly a series of mixed selection combat sorties together.
30 May–1 June	Marines of 1st Battalion, 3d Marines, support Afghan efforts to register voters for district community council elections through key leader engagements, messaging support, and security for election routes and facilities. More than 600 voters are registered with no security incidents.
7 June	The 1st Battalion, 9th Marines, relieves 2d Battalion, 3d Marines, in Nawa District.
22 June	President Obama announces a plan to withdraw 33,000 troops from Afghanistan by the summer of 2012. His administration faces continuous pressure from policy makers and the public to reduce the American presence in Afghanistan.
1 July	Marine Special Operations Team 8311 of Kilo Company, 3d Marine Special Operations Battalion, deploys to Afghanistan to conduct combat operations.
15 July	Marine Special Operations Golf Company (Rein) relieves Marine Special Operations Foxtrot Company in Farah Province and begins conducting village stability operations.
18 July	Gen David Petraeus hands over his duties as commander of U.S. and NATO troops in Afghanistan to Gen John R. Allen, USMC.
20–21 July	Elements of the 3d Reconnaissance Battalion and the 7th Engineer Support Battalion conduct Operation Street Without Joy II as a part of counterinsurgency efforts in Helmand Province.

1 August	Ramadan begins. Marines of 1st Battalion, 3d Marines, execute Operation Ramadan Mubarak, covering daytime patrols while ANSF troops increase nighttime patrolling. Additionally, the Marines continue peace and reconciliation mosque projects in each company area of operation (AO) and regularly join ANSF and Government of the Islamic Republic of Afghanistan (GIRoA) partners for Iftar dinners (concludes daily Ramadan fast).
7 August	The 1st Battalion, 6th Marines, relieves 2d Battalion, 8th Marines, in Helmand Province.
17 September	The 2d Battalion, 4th Marines, relieves 3d Battalion, 2d Marines, in Musa Qala and Now Zad Districts.
13 October	The 3d Battalion, 7th Marines, relieves 1st Battalion, 5th Marines, in Sangin District.
23 October	The 3d Combat Engineer Battalion (3d CEB) relieves 2d CEB in Helmand Province.
11 November	VMA-223 relieves VMA-513 at Kandahar Airfield.
15 November	The 1st LAR relieves 2d LAR in Reg-e Khan Neshin District, Helmand Province.
17 November	The 3d Battalion, 3d Marines, relieves 1st Battalion, 3d Marines, in Garmsir District.
19 November	The 2d Battalion, 11th Marines (-) (Rein), relieves 1st Battalion, 12th Marines (-) (Rein), to provide artillery support to the 1st Marine Division in Helmand Province.
25 November	Commandant of the Marine Corps, Gen James F. Amos, visits Marine Tactical Electronic Warfare Squadron 1 (VMAQ 1) at Bagram Airfield, Afghanistan, to announce that the squadron has been named the recipient of the 2011 Secretary of Defense Maintenance Award.
6–15 December	Marines of 1st LAR conduct Operation Eagles Hunt, an ANSF-led clearing and eradication action within the western portion of Reg-e Khan Neshin District.

2012

January–March	Elements of 2d Battalion, 6th Marines, support operations led by ANSF in Nawa District, Ghazni Province, and begin transferring positions from Coalition forces to ANSF command.
2–13 January	Marines of 2d Battalion, 4th Marines, conduct Operation Double Check, which clears the southern Musa Qala wadi area of Taliban influence and allows for the building of AUP posts throughout the AO.
15–28 January	Elements of 3d Battalion, 7th Marines, conduct Operation Dynamic Arch in the Sangin District, where Marines deny the enemy safe havens and expand ANSF influence in the area.
20 January	Video footage of Marines urinating on three Afghan corpses, believed to be those of Taliban insurgents, enrages Taliban fighters in Afghanistan and produces mixed reactions in the United States.
22–27 January	Elements of 3d Battalion, 3d Marines, support ANA forces during Operation Da Tunder Zeghelwol, a clearing action in Garmsir District.
26–31 January	Elements of 1st Reconnaissance Battalion (Fwd) conduct Operation Lion's Den I in Ghorak District, Kandahar Province, resulting in the capture of detainees, multiple insurgents killed and wounded, and the reduction of enemy weapons caches.
February	Marine Special Operations Team 8311 of Kilo Company, 3d Marine Special Operations Battalion (3d MSOB), returns to the United States from combat operations in Afghanistan.
1 February	The 1st Battalion, 8th Marines, relieves 1st Battalion, 6th Marines, in Helmand Province.
1 February–15 May	Elements of RCT-5 conduct Operation Harvest Moon alongside Afghan police forces in southern Helmand Province. This action continues to hinder the insurgent poppy cultivation and the spring harvest, dealing a severe financial blow to the insurgency. Afghanistan's National Interdiction Unit destroys more than 25,000 pounds of narcotics, confiscates numerous weapons, and detains more than 25 suspected traffickers across the battle space.

2 February	U.S. Defense Secretary Leon E. Panetta announces the Pentagon's plan to end combat missions in Afghanistan as early as mid-2013 and assign American troops to an advisory role until their withdrawal in 2014.
12–23 February	Elements of 1st Reconnaissance Battalion (Fwd) conduct Operation Lion's Den II in Kajaki District, resulting in the capture of detainees, multiple insurgents killed and wounded, and the reduction of enemy weapons caches.
15 February	Marine Special Operations Golf Company relieves Marine Special Operations Hotel Company in Herat City, Herat Province, and begins operations in nine different locations.
15 February–16 March	The 1st LAR (-) (Rein) conducts Operation Highland Thunder, a battalion-level operation intended to disrupt insurgent activity while expanding ANSF's presence westward across the battalion's AO.
17–19 February	Marines of 3d Battalion, 3d Marines, conduct Operation De Gham Payela alongside ANSF units. This operation is a series of coordinated actions meant to disrupt enemy operations in Garmsir District.
25 February	The 1st MarDiv (Fwd) relieves 2d MarDiv at Camp Leatherneck.
March–April	The 3d Battalion, 3d Marines, transfers a number of forward operating bases in Garmsir District to ANA control as a part of the Marines' mission to cede responsibility of the area to their Afghan counterparts.
4 March	Marines of 3d Battalion, 7th Marines, assist ANSF soldiers in redistributing their forces across the entire Sangin District. This action expands ANSF influence and provides security for the population of Sangin.
12 March	The II MEF (Fwd) transfers authority over RC-Southwest to I MEF (Fwd) at Camp Leatherneck.
17–23 March	Elements of 3d Battalion, 3d Marines, conduct Operation Dawn of Hope, in which four separate platoon-level aerial assaults disrupt enemy operations in the Safar and Durzay areas.
19 March	The 2d Battalion, 4th Marines, relieves 2d Battalion, 5th Marines, in Helmand Province.

28 March–5 May	Marines of 3d Battalion, 3d Marines, launch Operation Zema Pa Sar Ma Tera, a battalion-level action that successfully disrupts insurgent activity in Helmand Province through the use of aerial assaults, CAAT platoon operations, and various interdiction missions.
11 April	The 1st Battalion, 7th Marines, begins operations in the Sangin Valley.
23 April	The 1st CEB relieves 3d CEB in providing combat engineer support to Task Force Leatherneck.
May	The 1st Battalion, 8th Marines, focuses its efforts on the impending harvest season. A concentrated effort, Operation Blue Magic, is begun to prevent tax collectors, smugglers, and traffickers from influencing the local population. The Marines employ a biometric surge, increased night patrolling, and continuous disruption operations in the immediate area surrounding Kajaki District.
15 May	In accordance with the drawdown of forces in Helmand Province, 2d Battalion, 6th Marines, assumes responsibility over Garmsir District from 3d Battalion, 3d Marines. Also on this day, 3d LAR relieves 1st LAR, and Marine Light Attack Helicopter Squadron 469 (HMLA-469) relieves HMLA-369 at Camp Leatherneck.
16 May–15 June	Marines of 1st Battalion, 7th Marines, conduct company-size air assault operations throughout the Sangin Valley against Taliban insurgents.
17–19 May	Marines of 1st Battalion, 7th Marines, conduct Operation Sangin Moshtarak Naweed in the Sangin Valley. This action is the largest air assault operation in Operation Enduring Freedom (OEF) history, with more than 440 Marines and ANA soldiers participating and more than 20 Marine and Coalition force aircraft participating.
23 May–9 June	Operation Branding Iron is executed by 2d Battalion, 5th Marines (Rein), in the Gandumereze and Zamindawar areas of Helmand Province. This action engages the enemy's communication and logistics capabilities affecting the surrounding Musa Qala and Kajaki areas.
26 May	The 3d Battalion, 8th Marines, relieves 2d Battalion, 6th Marines, in Helmand Province.
22–27 June	Marines of 1st Battalion, 7th Marines, spearhead Operation Jaws D5, a multiorganizational air and ground operation to cordon and search Qaleh ye Gaz, Helmand

Province, which is a known hotbed for enemy activity. There are multiple small- and medium-arms firefights and several IED strikes that require seven air medical evacuations on the first day alone. After 1st Battalion, 7th Marines, counters unprecedented casualties with multiple aviation, armor, and artillery counterattacks, the assault Charlie and Delta Companies are extracted to friendly lines.

July — Marine Special Operations Team 8313 of Kilo Company, 3d MSOB, deploys to Afghanistan and attaches to Hotel Company, 2d MSOB, to conduct combat operations.

5–21 July — Elements of 2d Battalion, 5th Marines, conduct heliborne assaults in support of Operation Branding Iron II, which aims to disrupt the enemy's capabilities in the districts of Musa Qala and Kajaki.

17 July — Marines of 1st Battalion, 1st Marines, conduct Operation Doubledown, a partnered heliborne raid intended to disrupt insurgent operations throughout the battalion security area and allow ANSF forces to expand their area of influence in 1st Battalion, 1st Marines' area of operations.

15 August — HMH-361 relieves HMH-466 (-) (Rein) at Camp Bastion, Helmand Province.

30 July–2 September — Marines of 1st Battalion, 1st Marines, and 1st Battalion, 7th Marines, conduct operations in Khwaja Aziz, Helmand Province, as a part of the larger Operation Helmand Viper. Partnered heliborne raids disrupt insurgent operations in security areas, remove Taliban leaders and high-valued enemy targets from within the battalion's AO, and allow ANSF forces to expand their area of influence in the region.

September–December — Marine Special Operations Team 8313 of Kilo Company, 3d MSOB, returns to the United States from combat operations in Afghanistan.

14 September — Fifteen Taliban insurgents dressed in U.S. Army uniforms breach the perimeter at Camp Bastion. They attack Marine fixed- and rotary-wing aircraft with automatic weapons and rocket-propelled grenades (RPGs), destroying six McDonnell Douglas AV-8B Harrier II planes and badly damaging two others. The attack lasts several hours and ends with 14 insurgents killed and 1 captured by Camp Bastion's quick reaction force. Marines suffer two killed in action, one of them being LtCol Christopher K. Raible, commanding officer of VMA-211. This firefight marks

the first time the Marine aviators of VMA-211 have fought as infantry since the Battle of Wake Island in December 1941.

2 October	The 2d Battalion, 7th Marines, relieves 1st Battalion, 7th Marines, in Sangin District.
20 October	The 2d CEB relieves 1st CEB in Helmand Province.
November	Marine Special Operations Alpha Company relieves Marine Special Operations Charlie Company. Marine Special Operations Team 8321 of Kilo Company, 3d MSOB, also deploys to Afghanistan.
1 November	Combat Logistics Regiment 2 (CLR-2), CLR-15, and additional Army and Navy construction units begin Operation Golden Gate, which facilitates the construction of the Sangin Bridge over the Helmand River. Marines of 2d Battalion, 7th Marines, provide security support for the project.
18 November	HMLA-169 relieves HMLA-469 at Camp Bastion in Helmand Province.
2 December	The 3d Battalion, 9th Marines, conduct an aerial assault raid force mission in the Bari Desert, Helmand Province, to target a known narcotics lab. The mission concludes with one Marine killed in action and another wounded in action.

2013

1 January	The 3d Battalion, 9th Marines, supports Operation Tethys, a joint GIRoA, Helmand Provincial Reconstruction Team (PRT), and ISAF project to develop Afghan capacity to repair Helmand Province's canal system.
2 January	Operation Harakat Tah Cham-Too Shai (Prepare to March) commences with ANSF units in the lead and Marines of 3d Battalion, 9th Marines, and Weapons Company, Operational Reserve Force, RCT-7, in direct support.
15–18 January	Marine advisors of 2d Battalion, 7th Marines, accompany ANA soldiers into battle against enemy insurgents during Operation New Hope, which is meant to bring peace and stability to Kajaki, Helmand Province, and its surrounding areas.
26 January	CLR-2 relieves CLR-15 in Helmand Province, Afghanistan.
5–18 February	RCT-7 and CLR-2 conduct Operation Dynamic Partner, a retrograde action with the purpose of extracting equipment and personnel from Village Stability Platform

	Shurakay in Helmand Province to facilitate its demilitarization. VMA-231 flies in direct support.
19 February	HMH-461 relieves HMH-361 at Camp Bastion.
25 February	The 2d MarDiv (Fwd) transfers authority of Task Force Leatherneck to 1st MarDiv (Fwd) at Camp Leatherneck.
27 February–8 May	Marines of 2d Battalion, 9th Marines, support the Afghan National Interdiction Unit in Operation Psarlay Taba, which results in the seizure of more than 21,500 pounds of various forms of suspected opium in Marjah, Helmand Province.
28 February	The II MEF (Fwd) relieves I MEF (Fwd) at Camp Leatherneck.
1 April	The 2d Law Enforcement Battalion arrives in Afghanistan to begin patrol operations in the vicinity of Camp Leatherneck.
2 April	Elements of 3d Battalion, 4th Marines, begin turning over various forward operating bases to Afghan counterparts, following the scheduled relief in place and transfer of authority as dictated by the growing responsibility of Afghan troops for the security of their country.
21 April	The 3d CEB relieves 2d CEB in Helmand Province.
24 April	VMA-311 relieves VMA-231 at Camp Bastion.
28–31 April	ANSF leads Operation Aoqab se Hasht (38 Eagle) to push Taliban insurgents out of Sangin District.
May	RCT-7 conducts Operation Khaspeska (Nightmare) with the Afghan Territorial Force 444. This action sees aerial raids into known Taliban safe havens and command and control nodes in the Now Zad and Washir Districts of AO Tripoli.
12 May	VMU-3 relieves VMU-1 at Camp Leatherneck.
15–19 May	Operation Eagle Peak is conducted by 2d Battalion, 8th Marines, and elements of CLR-2, which enables the 6th Afghan Border Police (ABP) to provide medical treatment and evacuation in Taghaz, Southern Helmand, and allows the ABP to conduct operations against Taliban insurgents in the region.

18 May	HMLA-167 relieves HMLA-169 at Camp Bastion.
19–20 May	Operation Gridlock sees RCT-7 and ANSF patrols disrupt enemy freedom of movement in the vicinity of Ghorak, Kandahar Province.
23–27 May	Marines from 3d Battalion, 4th Marines, and 3d CEB, partnered with soldiers from the 42d Georgian Light Infantry Battalion, to conduct Operation Thunderstruck, which demilitarizes Patrol Base Lazika in the Musa Qala District, Helmand Province, and reinvests combat power within AO Tripoli.
24–25 May	Marines conduct Operation Blackmax for recovery of a Bell UH-1Y Venom helicopter from HMLA-167 that experienced a hard landing near FOB Robinson in Helmand Province.
18 June	NATO forces hand over the responsibility of Afghanistan's security to Afghan forces. U.S. forces still in-country shift from an advisory position to a focus on training Afghan troops and conducting special counterterrorism operations.
4 August	Marine Medium Tiltrotor Squadron 165 (VMM-165) relieves VMM-264 at Camp Bastion.
21 October	The 1st CEB relieves 3d CEB in Helmand Province.
16 November	VMU-2 relieves VMU-3 at Camp Leatherneck.
25 December	Gen Amos, Commandant; SgtMaj Micheal P. Barrett, sergeant major of the Marine Corps; Gen Joseph F. Dunford Jr., commander ISAF; and Medal of Honor recipient Sgt Dakota L. Meyer conduct a Christmas Day visit and serve a holiday meal to troops at Camp Leatherneck.

2014

January–May	Elements of 1st Battalion, 9th Marines, conduct operations of the Cameron Falls series in Nad Ali District, Helmand Province. The Cameron Falls series of operations is intended to disrupt Taliban threats against the Bastion/Leatherneck Complex.
4 February	The 2d MAW (Fwd) conducts a transfer of authority with 3d MAW (Fwd) at Camp Leatherneck.

5 February	The II MEF (Fwd) turns authority over RC-Southwest to Marine Expeditionary Brigade–Afghanistan at Camp Leatherneck.
17–19 February	Gen Amos and SgtMaj Barrett conduct a visit to Camp Leatherneck, where the Commandant discusses the future of the Marine Corps after OEF.
22 February	Operation Alcatraz is successful as Marines from 1st Battalion, 9th Marines, and soldiers of the ANSF detain four known Taliban low-level commanders and destroy an IED in Shingazak, Marjah District, Helmand Province. This strengthens the local populace's confidence in ANSF's ability to protect them.
15 March	The 1st Battalion, 7th Marines, relieves 3d Battalion, 7th Marines, in Helmand Province.
10 May	The 1st Battalion, 2d Marines, relieves 1st Battalion, 9th Marines, in Helmand Province.
15 May	HMLA-467 relieves HMLA-339 at Camp Leatherneck.
17 May	VMU-1, Alpha Detachment, relieves VMU-2 (Fwd) at Camp Leatherneck.
19 May	BGen Daniel D. Yoo, commanding general of RC-Southwest, visits VMU-1 at Camp Leatherneck to observe a Boeing Insitu RQ-21 Blackjack unmanned aerial vehicle demonstration flight and receive a brief on its capabilities.
27 May	President Barack Obama announces that U.S. combat operations for OEF will end in December 2014 and that all U.S. forces will be withdrawn from Afghanistan by the end of 2016.
5 April	Afghan presidential elections are held.
24 April	The 2d CEB relieves 1st CEB in Helmand Province.
14 June	Afghan presidential election runoff is held.
19 June	Operation Watkins, a battalion-level operation into southern Showal, begins. Conducted by 1st Battalion, 2d Marines, this action is meant to disrupt the flow of lethal aid throughout southern Showal by closing on the enemy and conducting interdiction missions along key avenues of approach.

26 October	The United States ends combat operations in Afghanistan after 13 years of fighting, handing over Camp Leatherneck to Afghan forces.
28 December	NATO officially ends combat operations in Afghanistan.

Appendix E:
Medal of Honor Citation

The President of the United States in the name of Congress takes pleasure
in presenting the
MEDAL OF HONOR to

Lance Corporal William Kyle Carpenter
United States Marine Corps

For service as set forth in the following

For conspicuous gallantry and intrepidity at the risk of his life above and beyond the call of duty while serving as an Automatic Rifleman with Company F, 2d Battalion, 9th Marines, Regimental Combat Team 1, 1st Marine Division (Forward), I Marine Expeditionary Force (Forward), in Helmand Province, Afghanistan, in support of Operation Enduring Freedom on 21 November 2010. Lance Corporal Carpenter was a member of a platoon-sized coalition force, comprised of two reinforced Marine rifle squads partnered with an Afghan National Army squad. The platoon had established Patrol Base Dakota two days earlier in a small village in the Marjah District in order to disrupt enemy activity and provide security for the local Afghan population. Lance Corporal Carpenter and a fellow Marine were manning a rooftop security position on the perimeter of Patrol Base Dakota when the enemy initiated a daylight attack with hand grenades, one of which landed inside their sandbagged position. Without hesitation and with complete disregard for his own safety, Lance Corporal Carpenter moved toward the grenade in an attempt to shield his fellow Marine from the deadly blast. When the grenade detonated, his body absorbed the brunt of the blast, severely wounding him, but saving the life of his fellow Marine. By his undaunted courage, bold fighting spirit, and unwavering devotion to duty in the face of almost certain death, Lance Corporal Carpenter reflected great credit upon himself and upheld the highest traditions of the Marine Corps and the United States Naval Service.

Barack H. Obama
President of the United States of America

Presidential Unit Citation

THE SECRETARY OF THE NAVY
The President of the United States takes pleasure in presenting the
PRESIDENTIAL UNIT CITATION to
MARINE EXPEDITIONARY BRIGADE-AFGHANISTAN
for service as set forth in the following
CITATION:

For outstanding performance in action against enemy forces from 29 May 2009 to 12 April 2010, in support of Operation ENDURING FREEDOM. Marine Expeditionary Brigade-Afghanistan conducted the most holistic counterinsurgency campaign since the Coalition presence in Afghanistan began in 2001. Operating in three separate and austere provinces that had been bereft of government efficacy for years, the Brigade constructed expeditionary bases and air fields, and struck decisively at the heart of the Taliban insurgency with Operation KHANJAR in July 2009. Cities and hamlets across the region, from Now Zad to Khan Neshin, resumed regional commerce and schooling for children, and participated in national elections. Concurrent with kinetic fighting, the Brigade engaged tribal, religious, and government leaders with population-centric civil-military operations that synchronized developmental efforts across 58,000 square miles of battle-space. In February 2010, Operation MOSHTARAK reclaimed Marjah, a strategic agricultural hub and narco-terrorist safe haven in the Helmand River Valley. Together with thousands of Afghan National Security Forces, the Brigade tangibly improved the geo-political landscape of Southwestern Afghanistan. By their outstanding courage, aggressive fighting spirit, and untiring devotion to duty, the officers, enlisted personnel, and civilian employees of Marine Expeditionary Brigade-Afghanistan reflected great credit upon themselves and upheld the highest traditions of the Marine Corps and the United States Naval Service.

For the President
Ray Mabus
Secretary of the Navy

Appendix F:
Selected Sources and Annotated Bibliography

(Selections in bold type appear in this anthology.)

Published Sources

Ansary, Tamim. *Games Without Rules: The Often Interrupted History of Afghanistan*. New York: PublicAffairs/Perseus Book Group, 2012.

Author provides a history of Afghanistan and argues that its major political, social, and cultural developments come from internal struggles rather than foreign invasions or interventions.

Ballard, John R., David W. Lamm, and John K. Wood. *From Kabul to Baghdad and Back: The U.S. at War in Afghanistan and Iraq*. Annapolis: Naval Institute Press, 2012.

Examines American strategy during the Afghan and Iraqi Wars, discussing how political decision making and poor counterinsurgency tactics hampered American efforts in both wars.

Blackard, Mark. *Killing Sheep: The Righteous Insurgent*. Kearney, NE: Morris Publishing, 2012.

The experiences of a former narcotics agent sent to Afghanistan to track terrorists, weapons builders, and drug smugglers are presented in this uncensored wartime thriller.

Bolger, Daniel P. *Why We Lost: A General's Inside Account of the Iraq and Afghanistan Wars*. New York: Houghton Mifflin Harcourt, 2014.

Drawing from past experiences as a troop commander in Iraq and Afghanistan, LtGen Bolger, USA, argues that the United States could have won its wars had it not been for faulty intelligence, decision making, and strategy and an inability to understand the enemy.

Chamblin, Joseph and Milo Afong. *Into Infamy: A Marine Sniper's War*. CreateSpace Independent Publishing Platform, 2015.

A memoir from one of the Marines charged in the Afghanistan urination video scandal.

Chandrasekaran, Rajiv. *Little America: The War within the War for Afghanistan*. New York: Alfred A. Knopf, 2012.

A *Washington Post* war reporter offers firsthand accounts of the Afghanistan War after

President Barack H. Obama's 2009 troop surge, noting that mistakes on every level—government, military, and societal—have failed Afghanistan.

Clement, Jeff. *The Lieutenant Don't Know: One Marine's Story of Warfare and Combat Logistics in Afghanistan.* Havertown, PA: Casemate, 2014.

Author remembers his service in Afghanistan as a lieutenant with Combat Logistics Battalion 6, leading support missions across dangerous territory and facing insurgents in battle.

Dyer, Lucas A. *A Battle Won by Handshakes: The Story of Alpha Company 1/5.* Bloomington, IN: iUniverse, 2014.

Author shares his experiences in Afghanistan, where he and his men realized that the best way to achieve success was to band together with their Afghan counterparts and work together to drive the Taliban back.

Fairweather, Jack. *The Good War: Why We Couldn't Win the War or the Peace in Afghanistan.* New York: Basic Books, 2014.

War correspondent uses interviews, archived materials, and personal experiences to detail the entire Afghan War (2001–14) and argues that the righteous intentions of American interventionism eclipsed what otherwise could have been a "good war." The United States can help to restore peace in Afghanistan only when it understands the country's limitations.

Franzak, LtCol Michael, USMC (Ret). *A Nightmare's Prayer: A Marine Harrier Pilot's War in Afghanistan.* New York: Threshold Editions, 2010.

Author provides a compelling story describing combat operations and individual experiences as executive officer of Marine Attack Squadron 513, "The Flying Nightmares," in Afghanistan from 2002 to 2003.

Folsom, LtCol Seth W. B., USMC. *Where Youth and Laughter Go: With "The Cutting Edge" in Afghanistan.* Annapolis: Naval Institute Press, 2015.

The commander of 3d Battalion, 7th Marines, describes his battalion's training and 2011 deployment to Sangin District in the Helmand Province, Afghanistan.

Gall, Carlotta. *The Wrong Enemy: America in Afghanistan, 2001–2014.* Boston: Mariner Books, 2014.

Embedded reporter argues that, because Americans failed to understand their enemy in the Global War on Terrorism, the Afghan people suffered the tragedy of war while Pakistan was truly to blame.

Golembesky, Michael, and John R. Bruning. *Level Zero Heroes: The Story of U.S. Marine Special Operations in Bala Murghab, Afghanistan.* New York: St. Martin's Press, 2014.

Following U.S. Marine Special Operations Team 8222, this book depicts a major operation to isolate and destroy the Taliban stronghold Bala Murghab.

Goodavage, Maria. *Top Dog: The Story of Marine Hero Lucca.* New York: Dutton Adult, 2014.

This title provides a wartime biography of canine Marine Corps hero Lucca during both tours in Afghanistan and Iraq. The author includes examples of battlefield bravery and the relationships that developed between Lucca and her Marine handlers.

Johnson, Robert. *The Afghan Way of War: How and Why They Fight.* New York: Oxford University Press, 2011.

This book examines the development of armed conflict in Afghanistan during the twentieth and early twenty-first centuries, including examples of resistance to political oppression, foreign influence, and armed intervention.

Kaplan, Fred. *The Insurgents: David Petraeus and the Plot to Change the American Way of War.* New York: Simon & Schuster, 2014.

Journalist describes how David Petraeus and a group of soldier-intellectuals revolutionized the U.S military for the twenty-first century, which had the unfortunate side effect of giving politicians more temptation to wage wars.

Kilcullen, David. *The Accidental Guerrilla: Fighting Small Wars in the Midst of a Big One.* New York: Oxford University Press, 2009.

The author offers a thoughtful cross-examination of current "small wars" in Iraq, Afghanistan, and other countries, and the overarching Global War on Terrorism. This book argues that America's contemporary foreign policies are blurring local struggles and global campaigns together and therefore hindering its ability to handle individual crises.

Kummer, Maj David W., USMC. *U.S. Marines in Afghanistan, 2001–2009: Anthology and Annotated Bibliography.* Quantico, VA: Marine Corps History Division, 2014.

This text offers an anthology of sources detailing U.S. Marine Corps operations and experiences in Afghanistan from 2001 to 2009.

Lowrey, Col Nathan S., USMCR. *U.S. Marines in Afghanistan, 2001–2002: From the Sea.* Washington, DC: Marine Corps History Division, 2011.

This monograph tells the story of Marine expeditionary operations in Afghanistan, its involvement in the Global War on Terrorism, and how Marines and sailors acquired a major role in Operation Enduring Freedom against Taliban and al-Qaeda forces.

Malkasian, Carter. *War Comes to Garmser: Thirty Years of Conflict on the Afghan Frontier.* New York: Oxford University Press, 2013.

Author offers an account of Garmsir, a small community in Helmand Province, and its experiences with the Taliban and NATO troops during the Afghanistan War.

McChrystal, Gen Stanley, USA (Ret), Tantum Collins, David Silverman, and Chris Fussell. *Team of Teams: New Rules of Engagement for a Complex World.* New York: Portfolio/Penguin, 2015.

This book discusses how Gen McChrystal restructured the Joint Special Operations Task Force to meet the threats of insurgent groups in the Middle East.

Moyar, Mark. *A Question of Command: Counterinsurgency from the Civil War to Iraq.* New Haven, CT: Yale University Press, 2010.

This source provides an analysis of American counterinsurgency strategies since the Civil War, offering case studies of successful plans that could serve as a model for current COIN operations.

Nordeen, Lon. *AV-8B Harrier II Units of Operation Enduring Freedom.* Oxford, UK: Osprey Publishing, 2014.

The third volume on USMC Harriers in combat, this book highlights the experiences of Harrier pilots in Afghanistan during Operation Enduring Freedom.

Obama, Barack H. "Remarks by President Obama in Address to the Nation from Afghanistan." Speech at Bagram Air Base, Afghanistan, 1 May 2012.

This presidential speech to the nation addresses the United States' efforts in Afghanistan and the strategies of the Obama administration designed to bring the war to an end.

———. "Remarks by the President to the Troops at Bagram Air Base, Afghanistan." Speech at Bagram Air Base, Afghanistan, 25 May 2014.

This presidential speech to troops stationed in Afghanistan acknowledges their role in preparing Afghan military forces to take the lead in securing their country and reiterates that America's war will be over by the end of the year.

———. **"Remarks by the President on the Way Forward in Afghanistan." Speech at the White House, Washington, DC, 22 June 2011.**

In light of recent successes of the troop surge in Afghanistan, this presidential speech to the nation announces a July deadline for a drawdown of forces.

———. **"Statement by the President on Afghanistan." Speech in the White House Rose Garden, Washington, DC, 27 May 2014.**

This presidential speech to the nation addresses the progress of America's drawdown and scheduled withdrawal of troops from Afghanistan.

Panetta, Leon. *Worthy Fights: A Memoir of Leadership in War and Peace.* With Jim Newton. New York: Penguin Books, 2014.

This source offers the professional and personal biography of Defense Secretary Leon E. Panetta and his experiences in political offices during the Global War On Terrorism, including his role as CIA director leading up to the death of Osama bin Laden.

Robinson, Linda. *One Hundred Victories: Special Ops and the Future of American Warfare.* New York: PublicAffairs, 2014.

Following special operations elements in Afghanistan during Operation Enduring Freedom, the author provides a testimony to the capabilities and limitations of these small, specialized units.

Sileo, Tom, and Col Tom Manion, USMC (Ret). *Brothers Forever: The Enduring Bond between a Marine and a Navy SEAL that Transcended Their Ultimate Sacrifice.* Boston: Da Capo Press, 2014.

This book tells the story of two brothers, one a Marine and the other a Navy SEAL, their personal experiences at the U.S. Naval Academy, and combat in Iraq and Afghanistan.

Streatfeild, Richard. *Honourable Warriors: Fighting the Taliban in Afghanistan—A Front-line Account of the British Army's Battle for Helmand.* South Yorkshire, UK: Pen and Sword Military, 2014.

The author, having served as a major with the British Army in Afghanistan, offers a personal account of day-to-day operations and combat against the Taliban in Helmand Province.

Walling, Michael G. *Enduring Freedom, Enduring Voices: US Operations in Afghanistan.* Oxford, UK: Osprey Publishing, 2015.

This book offers a collection of personal experiences of men and women serving in Afghanistan during Operation Enduring Freedom, supplementing oral histories with additional research on the Afghanistan War.

West, Bing. *One Million Steps: A Marine Platoon at War.* New York: Random House, 2014.

Embedded journalist provides the story of a rifle platoon during the closing days of the war in Afghanistan, as the Marines disregard an impractical strategy and instead continue the fight against a dangerous enemy.

———. *The Wrong War: Grit, Strategy, and the Way Out of Afghanistan.* New York: Penguin Random House, 2012.

Former Marine, assistant defense secretary, and embedded journalist Bing West offers a definitive history of the war in Afghanistan, proposing that the United States remove its troops, stop spending billions of dollars, and allow Afghan forces to fight for themselves.

Woodward, Bob. *Obama's Wars*. New York: Simon & Schuster, 2011.

This book provides a close look at the Obama administration during the war in Afghanistan, highlighting its decision making on issues of foreign policy and the complicated relationships between politicians, military officials, and civilian leaders on every level of government.

Zinni, Gen Tony, USMC (Ret), and Tony Koltz. *Before the First Shots Are Fired: How America Can Win or Lose Off the Battlefield*. New York: Palgrave-Macmillan, 2014.

The authors argue that America needs to reexamine its response to international crises and how it conducts military operations, drawing on history to support proposed alternatives.

Articles

"It's Not the Artillery's Fault." *Marine Corps Gazette* 97, no. 3 (March 2013): 40–43.

The author analyzes after action reports from Operation Enduring Freedom, arguing that rifle platoons need better organic weapons if they are to successfully combat insurgents in Afghanistan.

Avitt, Andrew S. "First Female Advisor Prepares for Afghanistan Deployment." *Leatherneck* 94, no. 2 (February 2011): 46–47.

This article proposes the advantages and disadvantages of deploying a female advisor to Afghanistan, as many Afghan men view women differently than their American counterparts.

Betik, Bart A. "Aviation in the Mountains." *Marine Corps Gazette* 94, no. 9 (September 2010): 41–43.

The author discusses how challenges presented by Afghanistan's mountainous and cold weather environment affect predeployment training and operations for Marine aviators.

Brill, Arthur P., Jr. "The Corps and the Way Ahead: General James F. Amos, Commandant of the Marine Corps." *Leatherneck* 95, no. 11 (November 2012): 16–20.

Gen Amos discusses the future of the Marine Corps in light of its withdrawal from Afghanistan and the end of Operation Enduring Freedom.

———. "Outcall with the CMC: Gen Amos Led the Corps during Tough Times on and off the Battlefield." *Leatherneck* 97, no. 10 (October 2014): 36–41.

Interviewer speaks with Gen Amos about his tenure as 35th Commandant of the Marine Corps before Amos cedes his post to Gen Joseph F. Dunford Jr.

Butler, LtCol Glen. "Wherever the President May Direct." *Marine Corps Gazette* 95, no. 11 (November 2011): 56–60.

This article offers commentary on the role of the Marine Corps as America's force-in-readiness and how that role will be influenced by operational changes, drawdowns, and budget cuts in the post-Operation Enduring Freedom era.

Cancian, 1stLt Matthew F. "Counterinsurgency as Cargo Cult." *Marine Corps Gazette* 97, no. 1 (January 2013): 50–52.

Author describes modern-day "cargo cults," as influenced by Marine Corps conventional operations and counterinsurgency doctrine in Afghanistan.

Carroll, LtCol Patrick, USMC (Ret), and Terry Walker. "Victory in Afghanistan." *Marine Corps Gazette* 94, no. 7 (July 2010): 22–25.

This editorial encourages Marines to teach Afghan police "critical combat skills" that go above what is taught by the Kabul Police Academy, so that the police have a clear advantage over insurgents while patrolling in southern Afghanistan.

Cranmer, 1stLt Valerie J. "Not Our Father's War." *Marine Corps Gazette* 99, no. 2 (February 2015): 87–90.

Accepting the shift from conventional to unconventional warfare, the author argues that Marines need to emphasize training for small wars, ethnic conflicts, and humanitarian operations during their predeployment periods.

Darack, Ed. "The Everyday Warfighters: Indefatigable Mission Fidelity in the Modern Toil of War." *Leatherneck* 94, no. 7 (July 2011): 32–35.

The embedded journalist follows a mobile Marine squad on its search for improvised explosive devices (IEDs) near a small patrol base in Afghanistan.

Davis, Dean. "'Charlie' Company Marines Travel Far and Wide." *Leatherneck* 93, no. 8 (August 2010): 14.

This article discusses the daily life of patrolling that goes on in Afghanistan, covered from the viewpoint of Charlie Company, 1st Light Armored Reconnaissance Battalion.

Dressler, Jeffrey. "Operation Moshtarak: Preparing for the Battle of Marjah." *Backgrounder* (11 February 2010).

This Institute for the Study of War publication analyzes the Battle for Marjah in Afghanistan, which was the largest Marine-Afghan joint operation since 2001.

———. "Operation Moshtarak: Taking and Holding Marjah." *Backgrounder* (2 March 2010).

In this second installment of "Operation Moshtarak," the author describes the course of the battle and its priority for security and governance in Helmand Province.

———. "Will the Marines Push Into Northern Helmand?" *Backgrounder* (10 June 2010).

This article debates whether American and British troops will use the momentum gained during Operation Moshtarak to move next into the districts of Kajaki and Sangin in northern Helmand Province.

Elfers, Maj Mark W. "Irregular Warfare and the Marine Corps." *Marine Corps Gazette* 97, no. 3 (March 2013): 53–55.

The author argues that Marines need to practice irregular warfare, which has replaced the Cold War as the most significant security challenge of the era, to successfully combat the Taliban in Afghanistan and terrorist groups worldwide.

Eikenberry, Karl W. "The Limits of Counterinsurgency Doctrine in Afghanistan: The Other Side of the COIN." *Foreign Affairs* 92, no. 5 (September/October 2013): 59–74.

In light of a troop drawdown, this article explores how the strategy of counterinsurgency re-emerged in America's wars against terrorism and offers opinions as to how the United States will continue its fight in spite of military withdrawals and budget reductions.

Galo, Ed. "Regimental Combat Team 6: 'India' Company, 3d Battalion, 7th Marines Builds Relations in Afghanistan." ***Leatherneck*** **95, no. 3 (March 2012): 32–33.**

Following a company of Marines in Afghanistan, the reporter discusses how conducting joint patrols with soldiers from the Afghan National Army (ANA) helps the Marines build positive relationships with allied troops and locals.

Garamone, Jim. "General Dunford Assumes ISAF Command in Kabul, Afghanistan." *Leatherneck* 96, no. 4 (April 2013): 32.

This article covers the change in command at the International Security Assistance Force (ISAF) in Afghanistan, as Gen Dunford assumes command from Gen John R. Allen, USMC.

Gordon, Thomas J., Jim G. Gruny, Michael L. Muller, William J. Nemeth, Wendell B. Leimbach Jr., and

Brendan M. Rodden. "Marine Armor in Afghanistan." *Marine Corps Gazette* 94, no. 10 (October 2010): 69–72.

These authors describe the use of tanks and other armored vehicles in irregular warfare and note that the M1A1 Abrams tank could be an enormous help to Marine air-ground task force commanders in Afghanistan.

Haas, Cody. "In Every Clime and Place: Marines Patrol and Provide Security around Camps Bastion and Leatherneck." *Leatherneck* 97, no. 11 (November 2014): 12.

This article follows a forward-deployed Marine infantry battalion as it concludes operations at Camps Bastion and Leatherneck before turning the area over to the Afghan National Security Forces.

Hayden, H. T. "Winning Hearts and Minds." *Marine Corps Gazette* 94, no. 6 (June 2010): 34–37.

President Obama authorizes a troop surge in Afghanistan, maintaining that the mission of American armed forces is to prevent the Taliban from overthrowing the Afghan government and to keep al-Qaeda out of the country.

Harkins, Gina. "After Action." *Marine Corps Times,* 1 April 2013.

A description of Sgt Robert W. Richards Article 32 hearing concerning the urination video.

Hlad, Jennifer. "Hearing Concludes for Marine Accused of Urinating on Taliban Bodies." *McClatchy-Tribune Business News* 19 March 2013.

This article describes the results of several hearings on the urination video as well as comments from the battalion commander.

Jernigan, LtCol Michael. "Otis Raible Defines Leadership." *Marine Corps Gazette* 97, no. 7 (July 2013): 13–15.

This article provides a description of the Marine response to the Taliban assault on Camp Bastion in 2012, which resulted in the destruction of several AV-8B Harriers and the deaths of LtCol Christopher "Otis" Raible and Sgt Bradley W. Atwell.

Johnson, Frances. "In Every Clime and Place: I MEF (Forward) Reactivates for Historic Afghanistan Deployment." *Leatherneck* 96, no. 12 (December 2013): 16.

The author covers the upcoming deployment of the I Marine Expeditionary Force (I MEF Forward) to Afghanistan's Helmand and Nimroz Provinces in Regional Command-Southwest (RC-Southwest).

———. "In Every Clime and Place: 'Tango' Btry Still Providing Artillery for Forces in Afghanistan." *Leatherneck* 97, no. 8 (August 2014): 14–15.

This article describes a composite artillery battery, composed of both M777A2 155mm howitzers and High Mobility Artillery Rocket Systems (HIMARS), operating in Afghanistan.

Johnson, Ned. "Camp Shorabak, Afghanistan: Radio Check: Marines Train Afghan Soldiers and Improve Communication Skills." *Leatherneck* 96, no. 9 (September 2013): 15–16.

Following a signals kandak, which is structured like a Marine Corps communications battalion, the author covers the actions of Afghan radio operators and data specialists who train with Marines to improve communication between units.

———. "In Every Clime and Place: Civil Affairs Marines Advise the Progressing Afghan Government." *Leatherneck* 96, no. 8 (August 2013): 20–21.

Civil Affairs Detachment, Regimental Combat Team 7 (RCT-7), contains 11 Marines who work with the RCT command and assist local government officials in Helmand Province, Afghanistan.

———. "In Time of Transition, Marines Support Afghan Advisors." *Leatherneck* 96, no. 7 (July 2013): 16–17.

This article documents a transitional period in which Marines turn from a role of conventional warfighters to that of advisors for Afghan National Security Forces.

———. "Last Marine Infantry Regiment Departs Helmand Province." *Leatherneck* 96, no. 10 (October 2013): 34–35.

This article covers the final action of RCT-7 as it prepares to leave Afghanistan, its author arguing that successful operations in Helmand Province allow RCT-7 to be the last unit of its size in the country.

Katz, David. "The Marine Corps in Afghanistan." *Marine Corps Gazette* 94, no. 6 (June 2010): 42–44.

The author comments on the current state of Marine operations in Afghanistan in the summer of 2010, noting that President Obama and American policy makers will face another big decision in terms of selecting a strategy for troops to implement against Taliban and al-Qaeda insurgents.

Lamothe, Dan. "Staff NCOs Face Courts-Martial in Video Urination Scandal." *Marine Corps Times*, 8 October 2012.

This article describes the charges presented against two staff sergeants in the urination video case.

Lombardo, Tony. "Dunford Hands Off Command in Afghanistan." *Marine Corps Times,* 26 August 2014.

This article covers the change in command at the International Security Assistance Force in Afghanistan, as Gen John F. Campbell, USA, assumes command from Gen Dunford.

Long, Austin. "In Every Clime and Place: Helmand Province, Afghanistan: Operation Pegasus II Nabs Bomb-Making Materials." *Leatherneck* 97, no. 1 (January 2014): 12–13.

Marines of 1st Battalion, 9th Marines, on a patrol to search for weapons caches and IED materials near Camps Bastion and Leatherneck.

Long, Austin, and Paul Peterson. "This Is My Life: Marines in Afghanistan." *Leatherneck* 97, no. 3 (March 2014): 36–40.

With the war in Afghanistan reaching its end, two journalists interview four Marines serving with 1st Battalion, 9th Marines, about their experiences in the Marine Corps and operations during numerous deployments to Afghanistan.

Lowrey, Col Nathan S., USMCR (Ret). "Operation Steel Dawn II: The Raid on Bahram Chah." *Marine Corps History* 1, no. 1 (Summer 2015): 6–22.

This article describes certain combat operations in Helmand, Afghanistan, conducted in October 2010.

Lubin, Andrew. " 'Arty' in Afghanistan–'We're Here to Shoot'." *Leatherneck* 93, no. 1 (January 2010): 38–42.

The Marine artillery battalion of 3d Battalion, 11th Marines, is tasked with supporting fire missions and providing mobile fire support in Helmand Province. This article describes its leadership, subordinate batteries, and achievements since deploying with Task Force Leatherneck.

———. "I MEF (Fwd): First into the Fight!." *Leatherneck* 93, no. 2 (February 2010): 24–26.

This article announces the forward deployment of the I Marine Expeditionary Force (I MEF) to Afghanistan, adding a sizeable force of infantry, artillery, reconnaissance, combat engineer, civil affairs, aviation, and support units to Afghanistan in augmentation of the 1st Marine Expeditionary Brigade.

———. "The Advisor Training Group: Making Afghanistan 'Transition' Successful." *Leatherneck* 95, no. 12 (December 2012): 26–28.

This piece describes the role of Marine advisor training groups and how their training prepares them to advise their Afghan counterparts in high-stress situations, such as ambushes, IED explosions, civil disobedience, and medical evacuations.

———. "Marine Artillery in Afghanistan: 1/11–Every Man an Infantryman and a Cannoneer." *Leatherneck* 94, no. 3 (March 2011): 38–41.

Following the individual batteries of 1st Battalion, 11th Marines, the author covers the many actions of Marine artillerymen in Afghanistan, including patrolling, running convoys, training local policemen, and advising local leaders.

———. "The New 'Strategic' Combat Correspondent in Afghanistan." *Leatherneck* 94, no. 9 (September 2011): 48–50.

This article describes the influence of new policies and modern technologies on embedded media coordinators, broadcast and press editors, combat cameramen, and public affairs specialists in Afghanistan.

Malkasian, Carter, and J. Kael Weston. "War Downsized: How to Accomplish More with Less." *Foreign Affairs* 91, no. 2 (2012).

With the United States withdrawing from Afghanistan, the authors propose an alternative to a costly strategy of counterinsurgency—counterterrorism—which relies heavily on special operations forces to conduct operations against terrorist leaders rather than seeing troops work alongside the Afghan people.

McNulty, Andrew J. "A Return to Instability." *Marine Corps Gazette* 97, no. 1 (January 2013): 53–56.

The author foresees a return to instability after the United States has withdrawn from Afghanistan, arguing that the Afghan people rely on American troops for security, economic development, and reconstruction after long years of Taliban rule.

Ostroska, Jessica L. "In Every Clime and Place: Southern Afghanistan: I MEF (Forward) Deploys to Afghanistan for the Last Time." *Leatherneck* 97, no. 3 (March 2014): 13.

The author describes I MEF's deployment to RC-Southwest for the last time in January 2014.

Owen, Jonathan, USMCR. "Border Security." *Marine Corps Gazette* 96, no. 4 (April 2012): 44–48.

This article comments on the failure of American counterinsurgency in Afghanistan after 10 years of fighting and blames this failure on poor border security, which otherwise would prevent terrorist insurgents from entering Afghanistan.

Peterson, Paul. "Armored Vehicle Drivers: Seeing the World through Their Eyes." *Leatherneck* 96, no. 11 (November 2013): 16–17.

LCpl Morgan Almazan describes her experiences as a motor vehicle operator with Combat Logistics Regiment (CLR-2) at Camp Leatherneck, acknowledging the closeness of her truck crew and the danger of their surroundings.

———. "Marine Turret Gunner in Afghanistan Says Job Keeps Him Going." *Leatherneck* 96, no. 12 (December 2013): 19.

LCpl Kevin Dunseith discusses his responsibilities as a turret gunner with CLR-2 in southwestern Afghanistan.

Rholes, Ryan. "MV-22 Osprey: Proving Itself in Combat." ***Leatherneck*** **93, no. 8 (August 2010): 42–45.**

This article describes the technological specifications of the Bell Boeing MV-22 Osprey and its operational history in Afghanistan with Marine Medium Tiltrotor Squadron 261 from 2009 to 2010.

Scanlan, Joseph. "In Every Clime and Place: Marines Confiscate Narcotics, Prevent Taliban from Reaping Profits." ***Leatherneck*** **97, no. 8 (August 2014): 15–16.**

This article follows the progress of Marine units tasked with locating and destroying caches of narcotics in Afghanistan that are a major source of Taliban funding.

Schoback, Capt Nicholas. "Warfighting: A Leadership Guide." *Marine Corps Gazette* 98, no. 2 (February 2014): 13–15.

The author explores the strategy of warfighting and the relationship between officers and their Marines, questioning how these elements will be affected as the Corps leaves Afghanistan and how a generation of Marines will influence a new era.

Seck, Hope Hodge. "Afghanistan Vet Pens Novel about War." *Marine Corps Times,* 1 March 2015.

The author interviews Silver Star recipient Capt Elliot Ackerman about writing a fictional novel on the Afghan War, *Green on Blue*, and why he chose to dedicate his work to two Afghan soldiers with whom he served in combat.

———. "After Base Handoff, Last Marines Depart Afghanistan." *Marine Corps Times*, 27 October 2014.

This article announces the transfer of Camp Leatherneck in Helmand Province from Marine control to that of Afghan National Security Forces, signifying the end of Marine operations in Afghanistan.

———. "His War Story." *Marine Corps Times*, 13 April 2015.

SSgt Joseph W. Chamblin discusses his plan to write a memoir following the Afghanistan urination case.

———. "'It's Your Medal': Read Kyle Carpenter's Inspiring Speech to Marjah Vets." ***Marine Corps Times*****, 16 February 2015.**

On the five-year anniversary of the beginning of Operation Moshtarak, Medal of Honor recipient Kyle Carpenter offered an encouraging speech to Marine veterans of the campaign.

———. "Marine 1-Star: Signs of Taliban Fractures in Afghanistan." *Marine Corps Times,* 16 January 2015.

BGen Daniel D. Yoo, who last commanded Marines in Afghanistan, offers a confident analysis of the ANA's ability to combat the Taliban and dissuades rumors that ISIS is recruiting men in Helmand Province.

———. "Marine Commanders Reflect on 13 Years of War in Afghanistan." *Marine Corps Times,* 5 November 2014.

The author speaks with six active-duty Marines about their thoughts on the war in Afghanistan. Although the Marines agree that American forces should withdraw from Afghanistan after 13 years of fighting, they are not confident in the ability of their Afghan counterparts to maintain security and improvement.

———. "MARSOC Corpsman to Receive Silver Star." *Marine Corps Times,* 19 March 2015.

HM1 Kevin D. Baskin, USN, is awarded the Silver Star for administering aid to a wounded Marine under enemy fire while attached to Marine Special Operations Team 8224 in South Zereko Valley, Afghanistan.

Vasgerdsian, Ed. "Portrait of a Wounded Warrior: The Story of Staff Sergeant Jason Pennock." *Leatherneck* 96, no. 8 (August 2013): 34–38.

This article follows a wounded Marine off the battlefield and into the hospitals for a number of corrective surgeries. SSgt Jason M. Pennock was wounded during a firefight in Helmand Province, where Taliban insurgents ambushed a team of ANA soldiers and Marine reconnaissance advisors.

Wilkes, John. "MALS-31 'Stingers' Support Operation Enduring Freedom." *Leatherneck* 96, no. 4 (April 2013): 17.

The author follows an all-volunteer detachment of Marine Aviation Logistics Squadron 31 (MALS-31) as it deploys to Helmand Province in support of Operation Enduring Freedom.

Yarbrough, Bobby J. "Gimlet-Eyed Unmanned Aerial Vehicles See It All for Marines." *Leatherneck* 96, no. 5 (May 2013): 13–15.

This article discusses the functions of the Marines' unmanned aerial vehicle (UAV) and describes how aerial surveillance, reconnaissance, and intelligence is crucial to maneuver elements on the ground.

———. "The Green Zone: Security and Stability Are Staples in Kajaki, a Former Taliban Stronghold." *Leatherneck* 96, no. 6 (June 2013): 8–9.

The journalist explores the relationship between Marine advisors and soldiers of the

ANA as the security of Kajaki, once a Taliban stronghold, becomes the ANA's responsibility.

Young, Joshua. "Last Marines Exit Sangin." *Leatherneck* 97, no. 7 (July 2014): 12–13.

This article depicts one example of the transition in Afghanistan, as Marine advisors leave Sangin to their Afghan counterparts. Forward Operating Base (FOB) Robinson is renamed FOB Rahatullah as it is placed under ANA command.

———. "Security, Voter Safety Top Priorities as Afghans Prepare for Elections." *Leatherneck* 97, no. 4 (April 2014): 10–11.

The combat correspondent reports on the results of a *shura* (a meeting between Afghan district leaders to discuss various issues) as Afghanistan prepares to transfer the power of its presidency for the very first time.

Index

— B —

— C —

— D —

— E —

— F —

— G —

— M —

— N —

— O —

— P —

— Q —

— R —

— S —

— T —

— U —

— V —

— W —

— Y —

d States
FL
er 2026

87976745R00171